Key Technologies for Tunnel Construction under Complex Geological and Environmental Conditions

Series Editor

Qihu Qian, Nanjing, China

This series is written on the basis of several difficult projects, aiming at complex geological conditions such as complex topography and landform types, complex geological structure and poor geotechnical engineering geology, systematically describes the Key Technologies for Tunnel Construction under Complex Geological and Environmental Conditions like mountains, underwater, soft and uneven strata, from the aspects of geological exploration, early warning and risk control, design, construction and digital application. It's a monograph that introduces innovative achievements and core technologies formed in the construction of complex geological tunnels in recent years. The series contains both texts and illustrations. It can be used as a reference for engineers, technicians, scientific researchers, teachers and students of related specialties in tunnels and underground engineering.

This series systematically describes the Key Technologies for Tunnel Construction under Complex Geological and Environmental Conditions like mountains, underwater, soft and uneven strata, from the aspects of geological exploration, early warning and risk control, design, construction and digital application.

Contents of the series are mainly from a number of key projects, such as Yuelongmen Tunnel of Chenglan Railway in Sichuan Province, Jinan Yellow River Tunnel Project, Qiyueshan Tunnel of West Hubei Expressway, Jiaozhou Bay Tunnel of Qingdao, Nanjing Yangtze River Tunnel and Wuhan River Crossing Tunnel, as well as the 973 Projects of the State-Mechanism, Prediction, Early Warning and Control of Major Water Inrush Disasters in Deep and Long Tunnels, Science and Technology Project by the Ministry of Transport "Study on Key Technologies of Underwater Large Section Shield Tunnel Construction under Complex Geological Conditions", National Defense Research Project "Three-Dimensional Advance Prediction Technology for National Defense Underground Engineering".

Rulu Wang · Fayun Liang · Jiaping Li · Zhao Yuan

Impact Analysis and Control Strategies of Overlying Loading/Unloading Effects on Soft Soil Shield Tunnels

SSTP

Springer

Rulu Wang
Shanghai Shentong Metro Co., Ltd.
Shanghai, China

Fayun Liang
Tongji University
Shanghai, China

Jiaping Li
Shanghai Shentong Metro Co., Ltd.
Shanghai, China

Zhao Yuan
SGIDI Engineering Consulting (Group) Co., Ltd.
Shanghai, China

Translated by
Yang Han
The University of Shanghai for Science and Technology
Shanghai, China

Juyan Wu
The University of Shanghai for Science and Technology
Shanghai, China

Shang Wu
The University of Shanghai for Science and Technology
Shanghai, China

Cheng Jiang
The University of Shanghai for Science and Technology
Shanghai, China

ISSN 2662-2904 ISSN 2662-2912 (electronic)
Key Technologies for Tunnel Construction under Complex Geological and Environmental Conditions
ISBN 978-981-95-1461-8 ISBN 978-981-95-1462-5 (eBook)
https://doi.org/10.1007/978-981-95-1462-5

Jointly published with Shanghai Scientific and Technical Publishers
The print edition is not for sale in China (Mainland). Customers from China (Mainland) please order the print book from: Shanghai Scientific and Technical Publishers.

This Springer imprint is published by the registered company Springer Nature Singapore Pte Ltd.
The registered company address is: 152 Beach Road, #21-01/04 Gateway East, Singapore 189721, Singapore

Foreword

Since the early 1950s, China has embarked on exploratory research into tunnel construction using the shield tunneling method. The completion of the Shanghai Metro test section in 1968 and the Dapu Road Tunnel crossing the Huangpu River in 1970 marked China's initial mastery of shield machine manufacturing and shield construction technology. By the 1990s, cities such as Shanghai and Guangzhou began extensively employing the shield tunneling method for metro tunnel construction. Particularly since the 21st century, over 40 cities across the country have initiated large-scale rail transit projects, with a significant proportion of tunnels built using shield tunneling. With rapid economic and urban development, various small- and medium-diameter pipeline projects, as well as large-diameter municipal road and railway tunnels crossing rivers, lakes, and seas, have seen vigorous growth, greatly advancing shield tunnel construction. As of September 2021, 49 cities in China have operational urban rail transit systems, with a total track length exceeding 8500 km and over 7000 km under construction.

Historically, some foreign cities experienced a surge in tunnel maintenance and remediation workloads due to overly rapid construction. Objectively speaking, high-intensity construction can lead to quality flaws, and the phenomenon of prioritizing construction over management remains widespread. Practitioners often overlook the impact of external environmental changes on metro tunnel structures, lack experience in tunnel maintenance and management, underestimate the hazards of ground loading/unloading, and lack follow-up measures for defect treatment. Consequently, multiple metro tunnels in various cities have suffered damage, necessitating emergency interventions. Thus, it is urgent to investigate the influence of loading/unloading on metro tunnels, deepening the understanding of its hazards in soft soil areas, and controlling its impact to ensure tunnel safety.

Against this backdrop, *Impact Analysis and Control Strategies of Overlying Loading/Unloading Effects on Soft Soil Shield Tunnels* was written. This book combines theoretical analysis, model testing, numerical simulation, and field observation, with Shanghai's typical strata as the research context. Drawing on over two decades of engineering experience in construction and operation management in

Shanghai, it systematically examines the effects of loading/unloading on the longitudinal and transverse deformation of shield tunnels, and establishes design and construction control standards of unloading/unloading. This work holds significant practical value for guiding loading/unloading operations above tunnels and ensuring tunnel safety.

The compilation team consists of experts with over 20 years of experience in metro construction, tunnel operation, maintenance, remediation, and theoretical research. Their profound theoretical knowledge, extensive practical experience, and remarkable research accomplishments are fully reflected in this book. I commend the team for their dedication and sincerely hope that this publication will contribute to ensuring the structural safety of shield tunnels in soft soil regions.

Shanghai, China
June 2022

Hehua Zhu

Preface

With accelerating urbanization and continuous population inflow, cities are expanding rapidly, leading to severe ground traffic congestion, air pollution, and reduced travel efficiency. Such "big city diseases" have reached a point where solutions are imperative, and rail transit has emerged as one of the most effective remedies. As of September 2021, 49 cities in China have operational rail transit systems, with over 50 more under construction—an unprecedented scale and speed of development. Cities like Beijing, Shanghai, Guangzhou, Shenzhen, Chengdu, Wuhan, and Nanjing now boast operational networks spanning 400–800 km, with rail transit accounting for 40%–70% of public transport ridership. Rail transit is playing an increasingly vital role in urban mobility.

However, such high-intensity construction also poses immense challenges for operational management. These challenges stem from insufficient accumulation of tunnel inspection technologies, equipment, and experience, as well as a lack of comprehensive management capabilities—from standard formulation to maintenance operations. Additionally, there is limited awareness of factors affecting tunnel structural safety.

Objectively, numerous factors influence tunnel structural integrity, including geological conditions, environmental factors, initial construction quality, vibrations and loads from train operations, ground landscaping, river filling and excavation (loading/unloading), adjacent construction activities, and maintenance quality. Among these, large-scale loading/unloading above tunnels has a particularly pronounced impact on tunnel deformation and damage.

In recent years, China has witnessed several severe cases of tunnel structural damage caused by engineering activities. Large-scale earthwork disposal, backfilling, and building construction have imposed substantial loads on tunnels, significantly altering the original strata and tunnel conditions. In mild cases, this leads to tunnel deformation and water leakage; in severe cases, tunnels must be shut down for reinforcement. The hazards of overloading above tunnels can be critical.

Generally, sedimentary strata are characterized by "soft-upper, hard-lower", and tunnel deformation patterns vary significantly depending on the strata. If the tunnel traverses homogeneous or mechanically similar strata, deformation primarily manifests as longitudinal settlement. If the strata at the tunnel's cross-section are mechanically weaker than those below, transverse convergence deformation dominates. As early as the 2000s, inspections of a section of Shanghai Metro Line 1 revealed significant deformation due to repeated ground loading/unloading. A comprehensive "external control, internal remediation" strategy—combining surface unloading and internal steel ring reinforcement—was implemented, stabilizing the tunnel structure.

Similarly, unloading above tunnels also threatens structural safety. Whether from river excavation or foundation pit construction, unloading disturbs stable strata and tunnels, altering stress states and causing segment deformation. The extent of impact depends on construction disturbance, with foundation pit excavations in soft soil being particularly hazardous and requiring stringent control measures. Though different in mechanism, unloading poses equally severe risks as loading.

Timely detection and correction of loading/unloading hazards can prevent accidents, but this approach entails high economic and time costs, along with elevated tunnel safety risks. Given past lessons, theoretical and practical research is essential for early detection, warning, and intervention.

This book employs theoretical analysis, model testing, and numerical simulation, focusing on Shanghai's typical strata, to examine the effects of ground loading and adjacent unloading on shield tunnel deformation. Using Timoshenko beam theory and the Pasternak foundation model, it establishes a longitudinal deformation calculation model for ground loading, considering factors such as load dimensions, eccentricity, and soil properties. Through laboratory model tests, it details preparation procedures and analyzes transverse deformation patterns under different strata conditions. Numerical simulations further elucidate transverse deformation laws under three typical Shanghai strata conditions and the time-dependent effects of unloading, providing theoretical guidance for tunnel safety.

Additionally, the book systematically presents field case studies on deformation patterns and remediation measures in soft soil tunnels under loading/unloading conditions. These cases, along with valuable engineering data, offer a solid theoretical foundation and practical insights for risk identification and mitigation, serving as a reference for researchers and practitioners.

Throughout the research and writing of this book, we have benefited immensely from the support of our colleagues at Tongji University and Shanghai Geotechnical Investigations and Design Institute Engineering Consulting (Group) Co., Ltd. Special thanks go to Qiang Yuan and Yanqi Fang for their computational analysis contributions, as well as Dr. Tingbo Wang, Dr. Ji Yin, Dr. Jingya Yan, Tao Tang, Ying Chen, Lujie Wang, and others for their assistance.

We are deeply grateful to **CAE Academician** Prof. Hehua Zhu for writing the Foreword and providing guidance throughout the compilation process. This book also cites design, construction, and research achievements from Shanghai Shentong Metro and other cities, as well as data from underground engineering and tunnel conferences and papers. We express our deep appreciation for these contributions.

Despite three years of diligent effort, limitations in the authors' expertise may have led to inadequacies. We sincerely welcome readers' critiques and suggestions.

Shanghai, China
June 2022

Rulu Wang
Fayun Liang
Jiaping Li
Zhao Yuan

Contents

Chapter 1
Introduction

1.1 Research Background

Since the reform and opening up, urban construction has developed rapidly, while the conflicts in urban transportation have become more and more prominent. Meanwhile, featuring safety, punctuality, large capacity, and high speed, rail transit has started to play an increasingly important role in broadening urban space, building a rapid and developed traffic network, and improving traffic environment.

Up to the end of 2020, 538 cities in 77 countries and regions around the world had established urban rail transit systems, covering a total distance of 33,346 km with more than 34,220 rail stations. Among them, China (excluding data from Hong Kong, Macao, and Taiwan) ranked first with a total operating mileage of 7,978.2 km, 23.9% of the world's total mileage. And Shanghai leads all cities with an operating mileage of 834.2 km.

From the perspective of construction history, China's urban rail transit development has accelerated over time. Figure 1.1 illustrates the annual mileage of rail transit infrastructure completed in China from 1978 to 2021. Since 2010, China's rail transit industry has been booming. Each year, the mileage of urban rail transit reaches record highs at an astonishing pace, increasing by 700–1,000 km annually after 5,000 km in 2017. In 2020, it exceeded 1000 km and ended at 1,241.99 km, a record high. According to the development plan, China's urban rail transit construction will maintain rapid development in the coming years. By the end of 2021,[1] Shanghai had built an 831-km urban rail transit network, with railway commuters accounting for over 67 percent of public transport passengers. Shanghai will develop a rail traffic network of over 1,000 km by the end of the 14th Five-Year Plan period with projects including Line 19 and the first phases of Lines 20, 21, and 23. Beijing

[1] All operational mileage data mentioned in this book are as of the end of 2021.

R. Wang et al., *Impact Analysis and Control Strategies of Overlying Loading/Unloading Effects on Soft Soil Shield Tunnels*, Key Technologies for Tunnel Construction under Complex Geological and Environmental Conditions,
https://doi.org/10.1007/978-981-95-1462-5_1

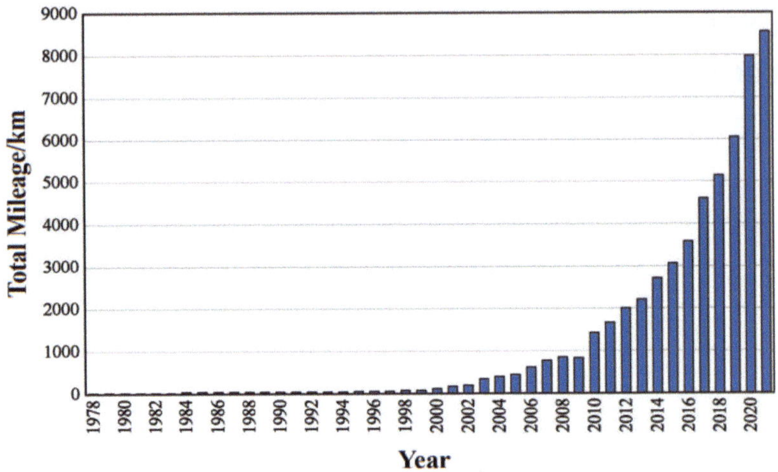

Fig. 1.1 Total mileage of urban rail transit built in mainland China (1978–2021)

has expanded its rail transit system to 783 km, with rail-based trips now making up 56% of public transport usage. Guangzhou currently operates 590 km of rail transit and plans to start 10 rail transit projects at a total length of 258.1 km in 2023. Shenzhen's operational rail network spans 431 km and is about to build 15 lines with a total length of about 570 km in 2022. Chengdu's rail transit system is developing rapidly. Its mileage reaches 558 km because the city had built 5 metro lines in December 2020. Chengdu will establish a rail transit network with a total length of more than 700 km by the end of 2024. Wuhan's current rail system measures 435 km and is ready to build 14 lines with a total length of 606 km by 2024.

The underground rail transit network has become essential infrastructure in modern cities due to its large capacity, good punctuality, and high speed. As these strengths naturally attract passengers to take subways, the areas along the metro lines also become prime locations for developing commercial residences, complexes, and other various facilities. However, for existing subway tunnels in equilibrium, all kinds of engineering activities around inevitably threaten their safe operation. When any tunnel accident occurs, it is difficult and costly to repair the tunnel which may even be irreparable and deserted. Therefore, all engineering activities around the subway must be strictly managed.

For the operational subway tunnels, common construction disturbance includes foundation pit excavation, new building construction, shield tunneling, pile construction, vegetation surcharge, or river excavation and unloading. Construction disturbance destroys the original equilibrium of the shield structure, causing stress redistribution in the tunnel which will experience longitudinal and transverse deformation to a certain degree. In severe cases, the disturbance will trigger some diseases, such as joint opening, segment cracking, or bolt failure, provoking extensive water leakage and even equilibrium failure. These problems endanger the normal operation of the tunnel. Therefore, it is necessary to strictly control the engineering activities around subway tunnels and conduct in-depth research on the influencing factors and deformation control over the activities.

(1) Surface loading accidents

Lin et al. (2000) stated that surface surcharge caused settlement in the No. 70 ring lining segment in the east of the 2# shaft of the Dapu Road Tunnel in Shanghai. After the opening to traffic of this tunnel section in 1970, the low-lying area was backfilled between 1973 and 1987, during which earth filling and concrete layer pavement were carried out four times. Each time, the load increment per unit floor area was 10–20 kPa, totaling about 70 kPa for the four times. The field tests showed an increase in settlement of 20–40 mm after each loading, with a total increment of 110 mm after all four loadings.

On January 18th, 2013, in South China, an unexpected large-scale spoil from nearby earthworks was piled up above and on both sides of a shield tunnel. With an average height of 4 m, and up to 6 m, the spoil led to a maximum differential settlement of 22.84 mm. On December 8th, 2018, after the site leveling of this area, up to 6 m of spoil was created again. The spoil caused a cumulative settlement of 27.8 mm, an ovality of 17‰, and numerous longitudinal through cracks in segments. Meanwhile, a large amount of damp spots, cracks, and significant misalignments were observed (Fig. 1.2).

On September 26th, 2009, backfilling construction of the Xiaolai River channel caused continuous differential settlement in the shield tunnel of Shanghai Metro Line 9 between Zhongchun Road and Jiuting Stations. By April 21st, 2010, seven months later, the settlement was up to 28 mm, and the radius of curvature in differential settlement was about 10,000 m (Fig. 1.3). Before river dredging, the overburden thickness above the tunnel was 9.5 m. On April 17th, 2010, the river channel was gradually excavated, reaching a depth of 5.5 m by May 16th, 2010, resulting in an unloading ratio (ratio of unloaded weight to total overburden pressure) of 0.58. The excessive short-term unloading of soil above the tunnel caused significant heave in the excavated section, with a maximum tunnel heave of 21.3 mm within one month (Fan et al. 2014).

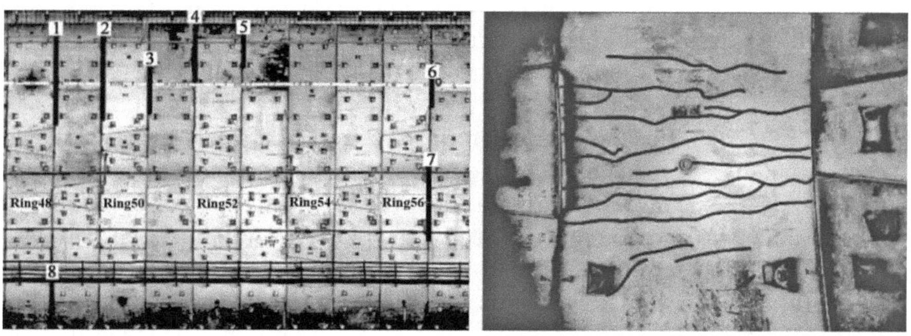

Fig. 1.2 Water leakage (left) and structural cracks (right) caused by loading

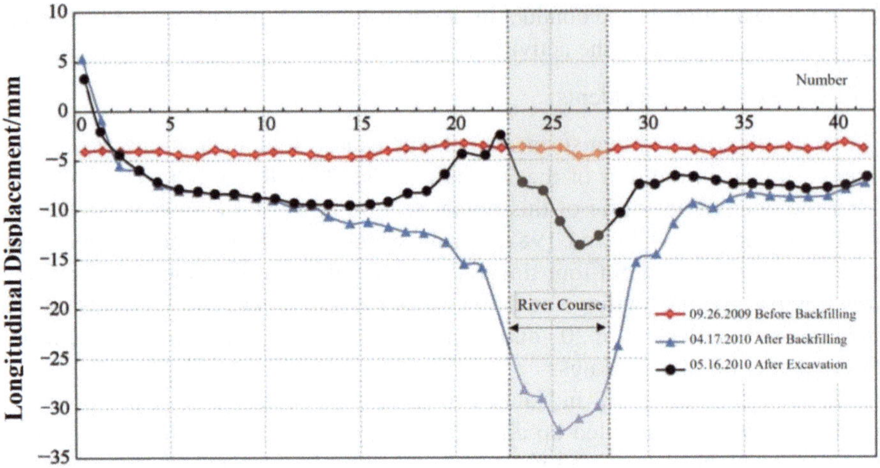

Fig. 1.3 Longitudinal deformation caused by river backfilling and excavation

In December 2008, a sudden large-scale soil piling occurred above a shield tunnel section of the Shanghai Metro. The situation caused water leakage and structural damage, which seriously threatened structural and operational safety. For instance, some of the lining rings experienced concrete spalling at the tunnel crown and bolt fractures. The longitudinal profile of this case is shown in Fig. 1.4 (Shao et al. 2016), where the piling affected Rings 280–600, with an average height of 4 m and a maximum localized pile height of 7 m. Within the loaded zone, the tunnel's horizontal diameter variation reached 14.5 cm, with the most deformed No. 576 Ring exhibiting a 15 cm deviation from its design diameter. The tunnel structure was ultimately restored by timely unloading, rectification, and structural reinforcement. Water leakage and structural damages, including joint opening, bolt fracture, and concrete shedding can be seen in Fig. 1.5 (Shao et al. 2016).

According to available statistics, there were dozens of unexpected soil surcharge incidents along metro lines across China in 2014. From 2004 to 2013, among reported accidents of Shanghai Metro caused by adjacent construction, there were eight cases where surface overloading triggered significant tunnel diseases.

(2) Unloading accident cases

Hwang et al. (2015) documented a deformation case in a shield tunnel in Taipei triggered by adjacent foundation pit excavation. The excavation was carried out in the vicinity of the existing tunnel to build a 4-storey basement for a 15-storey building. The excavation was carried out in six stages and the depth was 15.9 m. The pit was supported by diaphragm walls of 800 mm thick and 32.5 m long and by 5 layers of steel. To minimize wall deformation, the diaphragm wall technique was used to build a 600 mm-thick wall near the mid-span. The excavation was stopped as the convergence deformation exceeded the monitoring thresholds, so as to ensure tunnel

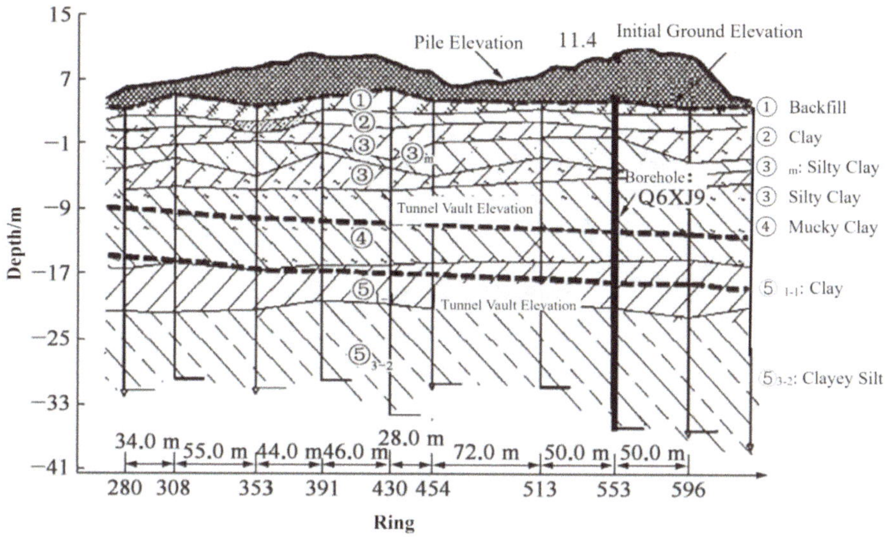

Fig. 1.4 Longitudinal profile of the loading above the tunnel

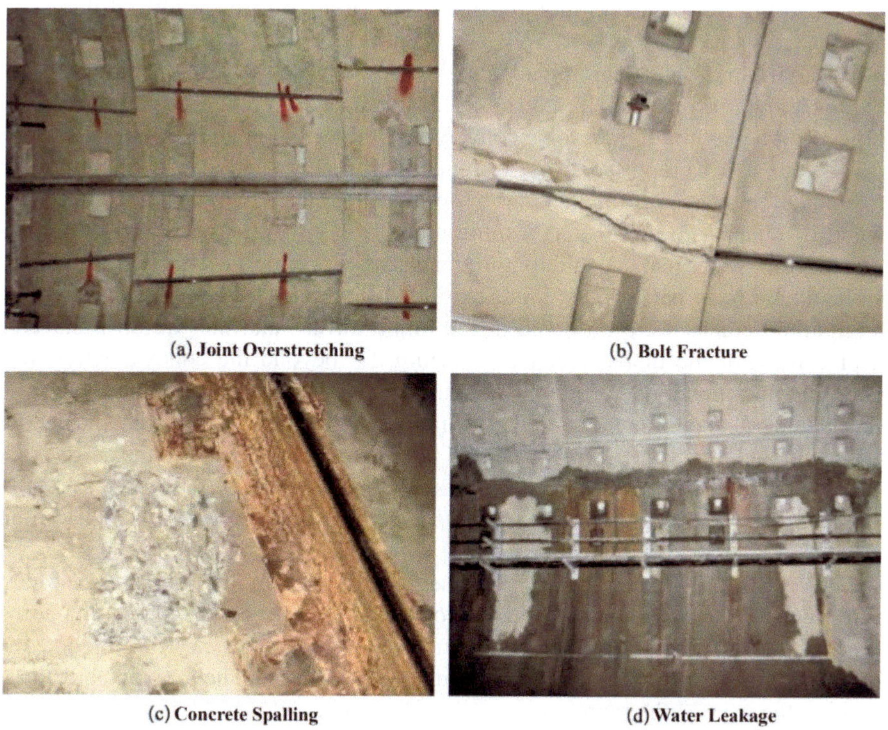

Fig. 1.5 Tunnel structural diseases caused by surcharge

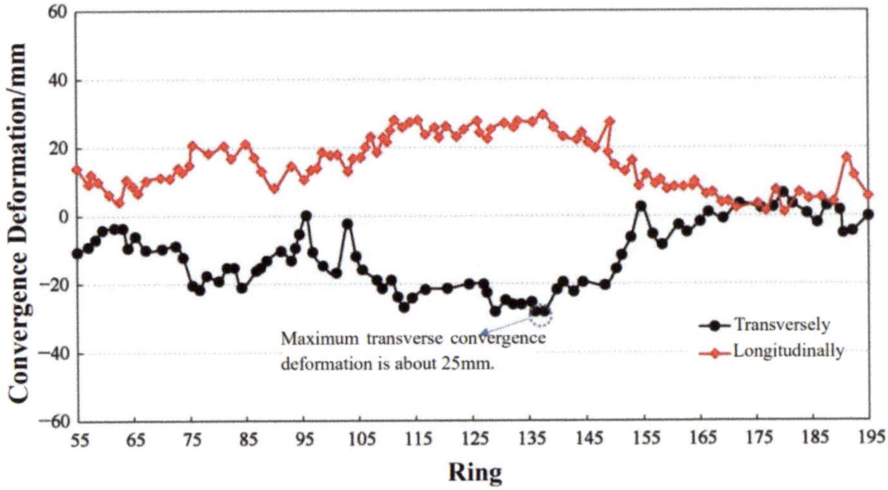

Fig. 1.6 Convergence deformation monitoring data between R55 and R195

safety and pursue long-term monitoring of the variation of the width of the R134 cracks.

The transverse and longitudinal tunnel convergence from Sep 9th to Dec 25th, 2009 is shown in Fig. 1.6. It shows the significant deformation between R55 and R150, with a maximum transverse convergence deformation up to 25 mm.

Lin et al. (2014) reported a case of tunnel deformation and displacement caused by foundation pit excavation in Ningbo Metro Line 1. The section from R60 to R260 which is adjacent to the pit has a minimum spacing of 7.15 m. Shortly after the construction, water leakage and longitudinal cracks occurred in this section. Meanwhile, the longitudinal segment joints exhibited significant widening. Subsequently, longitudinal cracks were observed at the tunnel invert, and the overall transverse displacement of the tunnel reached 33.6 mm, far beyond the control standard of ±10 mm, and the number continued to increase. A diachronic settlement curve of typical interval tunnel segments is shown in Fig. 1.7. The figure shows that the overall settlement and transverse displacement of the tunnel mainly occur in the earth excavation stage, during which the maximum settlement is −30.49 mm with a rate of −2.18 mm/d and the maximum transverse displacement 49.5 mm with that of 4.13 mm/d.

Chen et al. (2016) introduced that the segments in a section of the Ningbo Metro Line 1 were deformed and damaged due to foundation pit excavation and unloading nearby. Line 1 is the main line connecting the east and west city of Ningbo. The total length of the interval double-line tunnel is about 46.5 m, the horizontal distance between the axes of the tunnel is 12–15 m, and the crown burial depth is 9–15 m, with an average depth of 11.9 m. The inner and outer diameters of the tunnel lining are ϕ5.5 m and ϕ6.2 m respectively. The width of each lining ring is 1.2 m and its thickness is 35 cm. Each segment ring consists of 6 pieces of segment connected by curved bolts

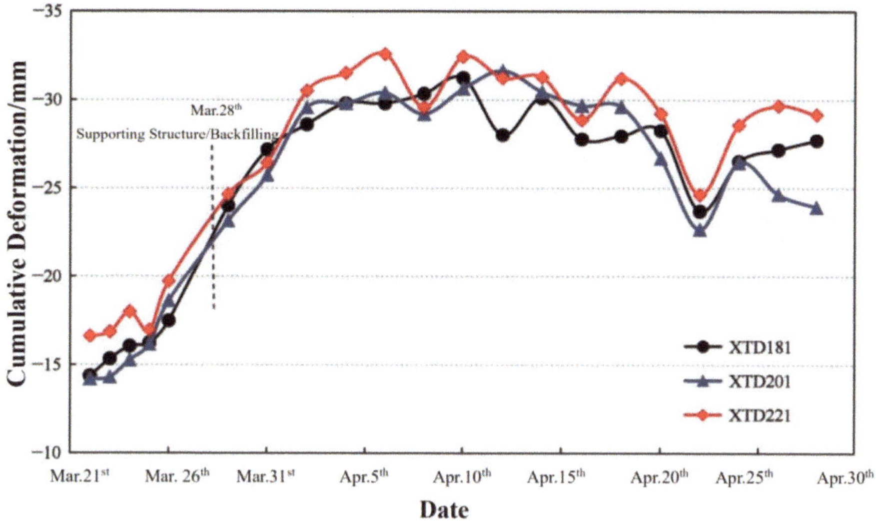

Fig. 1.7 Diachronic settlement curve of the left line of the interval tunnel in 2012

and the concrete strength grade reaches C50. The adjacent foundation pit is at the north of the double-line tunnel. Its retaining structure is supported by two reinforced concrete and bored piles, with a concrete strength of C30. During phase 3 of the pit excavation (to −11.4 m depth), longitudinal cracks with water leakage appeared at the invert of the left tunnel line. R40 to R414 of the left tunnel were damaged to different degrees, including increased misalignment, water leakage, segment cracking, and tail grease leakage. Meanwhile, the on-site monitoring data indicated that the left tunnel had undergone significant deformation and displacement. The maximum transverse displacement at R163 of the left tunnel is 33.5 mm, and the monitored transverse convergence, longitudinal convergence, and settlement at R221 are 21.9, 16, and 25.3 mm, respectively.

The cases above show that engineering activities, such as surcharge and foundation pit excavation (unloading) at the tunnel crown, may cause excessive deformation of adjacent tunnel structures, which leads to different degrees of damage to the tunnel in severe cases, especially in soft-soil areas. Therefore, it is significant to study the deformation mechanism of shield tunnels in soft-soil areas under surcharge and unloading, so as to ensure the stability and safety of subway systems during construction.

1.2 Research at Home and Abroad

1.2.1 Additional Stress and Displacement Fields Caused by Loading and Unloading

The two-stage method is widely used for analyzing the soil-tunnel interaction. It consists of two steps: calculate the additional stress and displacement fields of soil under surcharge; and calculate the soil-tunnel interaction, with the two fields as input conditions.

The calculation of additional stresses and displacements in soil is very important and complicated. To perform accurate and reasonable foundation settlement calculations, scholars from China and other countries have conducted in-depth research.

In the early research, the assumption that "the settlement of a point in the foundation depends only on the pressure acting at that point, and has no connection with the neighboring foundations" was commonly used to calculate the surface settlement under loading and unloading. This is the classical Winkler foundation model. The Winkler foundation model, however, overlooks the shear stresses in the foundation. It assumes that foundation deformation occurs only within the loaded area, and cannot be used to calculate the additional stress field in the soil.

Boussinesq assumed the soil to be a continuous, homogeneous, isotropic, semi-infinite, elastic medium and derived the additional stresses and displacements at any point in the soil under concentrated loading, which is the classic Boussinesq solution. Based on the Boussinesq solution, Mindlin obtained solutions for the additional stresses and displacement of the soil under internal loading in semi-infinite space, which is widely used in the calculation of the settlement of pile foundations.

Zhang et al. (2019a) and Liang et al. (2021a) introduced the Boussinesq solution into underground pipelines and tunnels, deriving the additional stress fields of underground pipelines and shield tunnels under surface surcharge. Liang et al. (2017) and Wei et al. (2020) proposed a calculation method for the additional stress field around tunnels induced by foundation pit excavation based on the Mindlin solution; Verruijt and Booker (1996), Loganathan and Poulos (1998) obtained an analytical algorithm for the soil displacement fields under tunnel excavation based on the theory of elasticity; Cheng et al. (2020) derived the displacement field induced by the unloading of pit excavation in the surrounding soils based on the sink–source method.

In fact, naturally deposited foundation soils exhibit a clear orientation during formation. But there are some differences that the soils feature anisotropy in the transverse and longitudinal directions. Transversely, they are approximately isotropic, while longitudinally, their behaviors are significantly different. Therefore, given such foundation soils that are naturally formed, it is more reasonable to use a layered foundation model to describe their state of stress and stress–strain relationship. All the above are based on the assumption of a homogeneous semi-infinite space soil which cannot account for soil stratification. When the tunnel is in layered soils with significant property variations, the Boussinesq or Mindlin solutions may yield substantial errors. In this context, it is necessary to investigate the additional stress and displacement fields in layered soils under surface surcharge.

According to the Code for Design of Foundation of China, the layer-wise summation method, which is practical for engineering applications, is introduced to incorporate the soil stratification and heterogeneity of the foundation into settlement calculations. When calculating the longitudinal deformation of a specific layer, this method substitutes the actual additional stress distribution in heterogeneous foundations with theoretical solutions derived for a homogeneous elastic half-space. However, the method is unsuitable for the calculation of the displacement field of soil under the large-area surcharge, because it is mainly for the calculation of the settlement of the center point of the foundation. Therefore, when a relatively large number of calculation factors are involved, the calculation becomes very cumbersome.

Bufler (1971) and Bahar (1972) proposed the transfer matrix method independently. Based on the Cayley-Hamilton theorem, they introduced transfer matrices for isotropic elastic layers in both 2D and 3D cases. However, due to the limited calculation conditions at that time, they failed to offer numerical solutions for stresses and displacements.

Booker and Small (1982) obtained the additional stresses and displacements in two-layer soil under surcharge by employing the finite element method (FEM) based on Biot's consolidation theory.

Wang (1982) derived expressions for the integration constants in the stress and displacement integral formulas of an N-layer elastic continuous system by solving the fundamental equations of elasticity theory using a step-by-step "recursive back-substitution method". By analyzing the expressions and the corresponding mathematical operations, a series of formulas suitable for stress and displacement integral calculations in N-layer systems were obtained. However, this method is cumbersome and inconvenient for further application. Based on the static equilibrium, constitutive, and geometrical equations, Zhong et al. (1992, 1995) proposed a transfer matrix equation of stress and displacement with respect to the partial differentials in longitudinal and radial coordinates. They used Hankel transform of this equation in radial coordinates to obtain the equation between the state vectors and the initial vectors of any layer. The state vectors can be expressed as the product of the transfer matrix and the initial vectors of the corresponding layer. Finally, the analytical solution in the Hankel transform domain can be obtained by solving the system of linear equations. The displacement and stress fields are then recovered through the inverse Hankel transform. This method proves advantageous for solving multi-layered elastic half-space problems, featuring clear physical concepts and mathematically concise formulations.

Based on elasticity theory and matrix theory, Zhong and Zhang (2003) derived the stiffness matrix for axisymmetric problems in a single-layer elastic half-space using Hankel transform and then assembled the global stiffness matrix by following the traditional FEM. The exact solution of the multi-layered elastic half-space axisymmetric problems under loading can be found by solving the algebraic equations constructed from this global stiffness matrix. Ai and Wu (2008) proposed a method for constructing the total stiffness matrix of stratified soils using "analytic layer elements". There is no overflow or ill-conditioned matrix during calculation because the elements in the stiffness matrix contain only negative exponential terms.

Zhang et al. (2009) and Liang et al. (2020) respectively studied the additional stresses in layered foundations under tunnel excavation unloading and surface surcharge employing elastic layered half-space theory.

1.2.2 Soil-Tunnel Interaction Analysis

After obtaining the additional stress or displacement input of the tunnel structure, many scholars have carried out extensive research on the soil-tunnel interaction to determine the internal force distribution and displacement response and to analyze the force and deformation characteristics in soils.

Winkler, two-parameter Pasternak, and three-parameter Kerr models, etc. are often used for soft soils. Among them, the Winkler foundation model is most widely used with its clear concept and quick calculation, as it regards soil as a series of mutually independent soil springs. However, the deviation between calculated and practical results is often large. To address this, the Pasternak model greatly reduces the calculation deviation by taking into account the shear stiffness of the soil. The three-parameter Kerr foundation model has the highest calculation accuracy, but its calculation process is more complex. The Euler–Bernoulli and Timoshenko beam assumptions are often used for tunnel structures. The former considers only the effect of bending, while the latter can calculate the amount of tunnel misalignment by integrating the shear deformation.

Jiang et al. (1999) conducted pioneering mechanical analysis of the longitudinal deformation of the tunnel caused by surface surcharge, and proposed to use the Boussinesq formula to calculate the additional stress at the tunnel crown and to obtain the stress distribution of the longitudinal tunnel sections based on the theory of elastic foundation beams. However, there is no specific research. Dai et al. (2006) used the Boussinesq solution to calculate the additional stresses in the soil layer caused by the surface surcharge, assuming that the longitudinal bending of the tunnel conforms to the Euler–Bernoulli beam theory. However, their study did not address segment misalignment issues.

Wang et al. (2005) regarded the tunnel structure as an elastic foundation beam in the soil layer. They used the Boussinesq solution to calculate the additional stresses in the soil layer caused by loading/unloading, and calculated the deformation of the tunnel structure based on the Winkler foundation model. Li et al. (2011) established a longitudinal internal force model of shield tunnel structure under surface loading based on the elastic foundation beam theory, so as to calculate the additional settlement and internal force generated in the tunnel. In this model, the reduction in bending stiffness caused by joints was considered.

Huang et al. (2012) calculated the tunnel response caused by unloading using Winkler, Pasternak, and Kerr foundation models, respectively. They found out that the Kerr model is more accurate compared to the Winkler and Pasternak models. Qu et al. (2016) equated the deeply-buried shield tunnel as a double elastic foundation beam. They used the Boussinesq solution to derive the additional stresses in the upper part of the tunnel, and finally adopted the FEM to obtain the longitudinal displacements and internal forces of the tunnel, and compared them with the values from the Winkler foundation model. It showed that the calculation results from the double elastic foundation beam are significantly smaller than those from the Winkler model.

Liang et al. (2017) studied the effect of foundation pit excavation on the deformation of underlying tunnels using the Timoshenko beam theory and the Mindlin solution. Gao (2017) proposed an analytical formula for the longitudinal displacement of crossover tunnels based on the Boussinesq and Mindlin solution, and analyzed their deformation response under surface surcharge. Wei et al. (2018) analyzed the effects of surcharge magnitude and eccentric distance on tunnel segment misalignment, longitudinal settlement, and transverse deformation based on energy variation methods and a "shear dislocation model". Kang et al. (2018) used the Boussinesq solution to obtain the additional stress above the tunnel, and studied the longitudinal deformation response of tunnels under different surcharge magnitudes and burial depth. Zhang et al. (2019b) assumed that the foundations are consistent with the Kerr model and simulated tunnels based on the Timoshenko theory to study the impact of newly-built tunnels on existing tunnels. Zhang et al. (2019a) analyzed the bending moments and transverse displacements of pipelines under these two input conditions based on the Boussinesq solution. They compared the calculation results with the FEM solutions and found out that the results obtained from the displacement field were more reasonable. Wei et al. (2020) obtained the transverse surcharge load distribution of tunnels under foundation pit excavation based on the Mindlin solution. They took into account the redistribution of internal forces during the coordination of tunnel displacement and discussed the influence of transverse deformation on the stress release coefficients of the foundation pit sidewalls and the tunnel burial depth.

1.2.3 Numerical Analysis Methods

Scholars in China and overseas usually adopt two methods in numerical simulation. One is to build a plane strain model based on pile unloading and tunnel structure, which is easier for modeling and calculation but has a relatively rough result. The other is to construct a 3D numerical model. In such a high-precision model, the structural details of the tunnel structure along with the interaction between soil and tunnel must be taken into account. While the modeling process of the second method is more complex, the results better reflect the actual engineering conditions.

Qi et al. (2005) applied the Mohr–Coulomb elastoplastic constitutive model into the 2D FEM of foundation pit excavation. They compared the calculation results with the measured data, and it showed that the numerical simulation was consistent with the engineering monitoring data, which verified the numerical modeling. Ye (2007) used ABAQUS to establish a 2D model to simulate the settlement of existing shield tunnels under four surface surcharge conditions of 5, 10, 20, and 40 kPa. It showed that the tunnel settlement is linear with the surcharge amount. And the tunnel settlement is closely related to the compression modulus of the underlying soil layer, that is, the larger the compression modulus is, the smaller the settlement is.

Zheng et al. (2016) build a 2D plane strain model based on PLAXIS finite element software with the small-strain hardening (HSS) model to simulate the soil. They calculated the effects of different support forms and the maximum transverse displacement on the deformation areas of the tunnel but failed to consider the structural details of the tunnel structure along with the changes in soil properties. Zhang et al. (2011) established a 3D finite element model, presented the unloading effect of foundation pit excavation by using the Element Birth and Death technique, and simulated the tunnel with a simple beam unit. During this process, they took into consideration the effect of different factors, such as buried depth, soil properties, and outer diameter of the tunnel, on longitudinal deformation. The results showed that the finite element calculation results were generally consistent with the theoretical solutions obtained using the Mindlin basic solution and the Winkler model. Wu et al. (2012) established a 3D MIDAS/GTS model based on a subway-adjacent excavation project in Guangzhou. They analyzed the influence of bilateral deep foundation pit excavation on operational subway tunnels by simulating different excavation sequences and comparing results with the monitoring data. The findings indicated that compared to asymmetric excavation, symmetric two-sided excavation of the foundation pits created less transverse displacement response in the transverse tunnel, which, however, is unfavorable for longitudinal displacement control.

Ge et al. (2013) used the district modified stiffness method (DMSM) to study the stiffness degradation at the segment joints, which provides a simplified simulation method for the high-refinement modeling of tunnels. Peng et al. (2013) used a shell-joint model to simulate the connection between the longitudinal and circumferential directions of subway segments, which is not convenient for practical engineering applications since the parameters of the shell-joint model need to be determined based on experimental results and involve subroutine programming. Su and He (2015) established 3D numerical computational simulation for longitudinal deformation analysis by using a shell element to simulate the tunnel segments, a spring element to simulate the contact between segments as well as the tensile stresses of the bolts.

Lu et al. (2017) developed 2D segments and 3D refined segments, both of which are joint discontinuous models, by simulating the ground reaction with a circumference unidirectional compression spring. They summarized the influence patterns of loading and unloading on the deformation of shield tunnel segments and on the change of the material damage in three kinds of soil conditions, and obtained safe limits for upper surcharge of shield tunnels in medium clays, consolidated clays, and stiff clays. Sun (2018) constructed a 3D FE model by adopting the structural parameters from Hangzhou Metro Line 1 and using MIDAS/GTS, so as to study the influence of different surcharge locations and burial depth on the tunnel deformation. Research showed that under the same surcharge condition, as the burial depth increases, the additional crown displacement decreases more significantly than haunch displacement. And the surcharge migration will decrease the displacement and deformation of tunnels.

Through numerical simulation, Zhang et al. (2020) studied the misalignment and damage characteristics of circumstantial joints in sequential, reverse, and tangential shear. They also analyzed the influence of inclined bolts and mortise-tenon structures on the shearing property of circumferential joints and proposed a safety index for joint deformation based on the results from tests and numerical simulations. Sang et al. (2020) elucidated the bearing stages of different circumstantial joints and the nonlinear laws of tensile and shear stiffnesses of the joints. Based on this, they adopted a 3D solid nonlinear spring model to analyze the influence of surface surcharge on the deformation parameters of tunnels. Ying et al. (2021) used numerical analysis to study the effect of different pit excavation conditions on tunnel deformation.

When tunnels and foundation pits are in parallel alignment, the maximum tunnel displacement is 1.6 times higher than that in perpendicular alignment. Increasing the longitudinal bending stiffness of the tunnel can effectively reduce the maximum displacement, but this mitigation effect diminishes with increasing tunnel-excavation spacing; longitudinal excavations affect the displacement and heave range of tunnels, while transverse excavations mainly influence the displacement.

Surface loading and unloading will lead to excessive convergence, joint damage, or concrete spalling, all of which threaten the structural safety or normal operation of the tunnel structure. When such structural issues occur, it is necessary to fix structural damage through rehabilitation methods and conduct rectification through deformation treatment. Internal steel-plate reinforcement and grouting rectification are commonly used methods currently.

Zhang et al. (2014) used numerical simulation to analyze the effect of lateral grouting on the transverse deformation of shield tunnels. The research indicated that when the grouting quantity is small, the joint opening can be significantly reduced but the convergence reduction is not significant; when the grouting quantity is large, grouting will cause some joint misalignment and lateral tunnel movement. Zhou (2015) studied the difference in bearing capacity and damage mechanism of longitudinal joints before and after plate reinforcement by establishing a 3D numerical model of longitudinal joints.

Zhao et al. (2016) investigated the effects of steel plate reinforcement timing, bond strength of epoxy resin, and steel plate size on the overall structural performance of shield tunnel lining by numerical modeling. It showed that improving the bond strength of epoxy resin has the most significant effect on enhancing the ultimate bearing capacity of composite lining; it is more effective to reinforce the tunnel segments when they have been deformed while the plastic deformation is small. It is also recommended that the thickness of the steel plate should exceed 20 mm. Guo (2016) used FEM to study the effect of grouting uplift for treating differential settlement of shield tunnels in soft soil and found that grouting has the most significant uplift effect on tunnels in upper-soft and lower-hard strata, while the least effective in upper-hard and lower-soft strata.

Qiu et al. (2018) studied how to use high-pressure vertical jet grouting piles for regulating differential settlement based on the FEM, which greatly improved the stability of operating tunnels. Based on on-site measurements and numerical simulation, Zhang et al. (2019c) studied the effect of grouting on reducing tunnel deformation. The results showed that the tunnel convergence can be reduced by 25% through two-sided grouting, which significantly reduces joint opening and improves the waterproof performance of the tunnel. They recommended that the spacing of grouting holes and the grouting sequence should be strictly controlled.

Zhang et al. (2019d) proposed a framework for the steel-plate reinforcement design of shield tunnel segments. They simulated this method for deformed segments through the 2D FMM, which proved that the reinforcement of steel plates in tunnels based on this method can effectively reduce the cost while ensuring the reinforcement effect.

Numerical simulation and theoretical calculation results are of guiding significance, which can reflect the influence of loading/unloading conditions on tunnel structures. To obtain more practically reliable results, on-site or indoor tests are required for verifying their validity.

1.2.4 Model Tests

Environmental impact assessment (EIA) and on-site monitoring are generally required when surcharge and unloading are conducted around operating tunnels. Displacement monitoring points in existing operational tunnels provide on-site data. However, due to cost and practical conditions, on-site monitoring is often limited in acquiring bending moment and earth pressure changes of the tunnel structure. Therefore, laboratory model tests are important for studying the influence of surcharge and unloading on adjacent tunnels.

Byun et al. (2006) conducted scaled model tests to simulate the effect of foundation pit excavation on existing tunnels, and measured the forces on lining segments and heave deformation caused by excavation. Wu and Du (2014) investigated the deformation characteristics of the tunnel under the conditions of tunnel burial depth and loading position through laboratory model tests. The burial depth is set at $1D$, $2D$, and $3D$ (D is the tunnel diameter), and the loading position is set at the center, $1D$, $2D$, and $3D$. The tests showed that: under the same surcharge, tunnels with burial depths $\leq 2D$ and loading offsets $\leq 1D$ are most significantly affected.

Huang et al. (2018) used a grooved model to simulate the reduction in bending stiffness at joints and established a comprehensive design method for tunnel grooved models. On this basis, to study the changes in soil pressure around the tunnel caused by surface surcharge, Zhang et al. (2016) carried out a 1:10 scale laboratory similarity model test. The test model box and grooved segment models are shown in Fig. 1.8. Loading thickness controlled the load magnitude. Figure 1.9 depicts the experimental curves of tunnel deformation under surcharge, and it shows that under the surface surcharge, the tunnel segments face lower soil pressure from the transverse side than vertical pressure directly above. The longitudinal and transverse convergence deformation gradually increases as the surcharge thickness (in multiples of the tunnel diameter *D*) grows. Through the analysis of the soil pressure data, it is concluded that the longitudinal soil pressure caused by surface surcharge in the soft soil exceeds that calculated from the soil column theory.

Based on the foundation pit evacuation near Xi'an Metro Line 4, Li (2017) carried out centrifuge tests to analyze the effect of deep foundation excavation on tunnel structures. It showed that during foundation pit evacuation, the tunnel is more vulnerable to the stress from the longitudinal side than from the transverse side. The stress

(a) Test Box Model (b) Grooved Segment Models

Fig. 1.8 Model test setup and slotted segment model

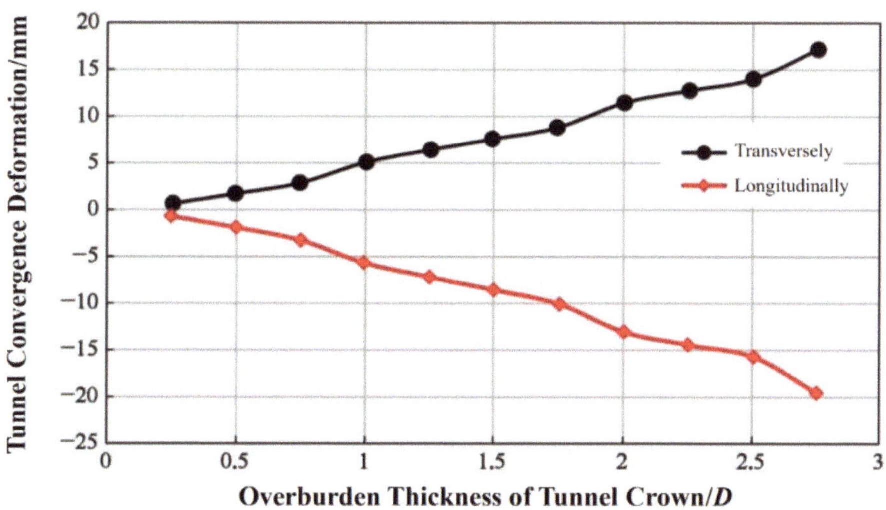

Fig. 1.9 Test result

is overall unloading. The excavation will lead to the situation where the tunnel inflection points are likely to turn counter-clockwise along the longitudinal axis. Gao et al. (2018) combined laboratory model tests and numerical simulation. They used PVC pipes to investigate the deformation response of crossover tunnels under surcharge. The test results agreed with the simulation results. It showed that under surcharge, the crossover of tunnels causes a decrease in the displacement of the upper tunnel and an increase in that of the lower tunnel. Especially, the interaction between the two crossover tunnels of small angles is more pronounced, which should be paid attention to.

Zhang et al. (2018) focused on the adverse effects of unloading–reloading action caused by asymmetric foundation pit evacuation on the constructed subway tunnels. They carried out geotechnical centrifuge tests to analyze the displacements and force conditions of the existing tunnels in different excavation steps and under various loads. The test similarity ratio is 1:60, a total of three excavations. They measured the tunnel bending moment and the size of the surrounding soil pressure during excavation. The foundation pit excavation layout and cross-section measurement point distribution are shown in Fig. 1.10, and the arrangement of the centrifuge test in the excavation is similar. The tests showed that the asymmetric unloading of the foundation pits leads to tunnel heave and offset, so the secondary excavation disturbance is more significant; the disturbance near the centerline of the foundation pit is more obvious, and the free surface of the foundation pit is influenced apparently.

Liang et al. (2021b) used a scaled model test to investigate the hardness conditions of the soil stratification through which the tunnel passes, and the effect of surcharge location on the lateral deformation properties of the segments. The measurements included the change curves of the ovality, joint opening, and the additional earth pressure in the graded loading/unloading. The tests show that the additional earth pressure in the hard soil layer is more evenly distributed than that in the soft soil layer, and the pressure in the soft soil layer is mainly distributed in the transverse direction, and the pressure decreases gradually as the distance between the surcharge and the tunnel increases.

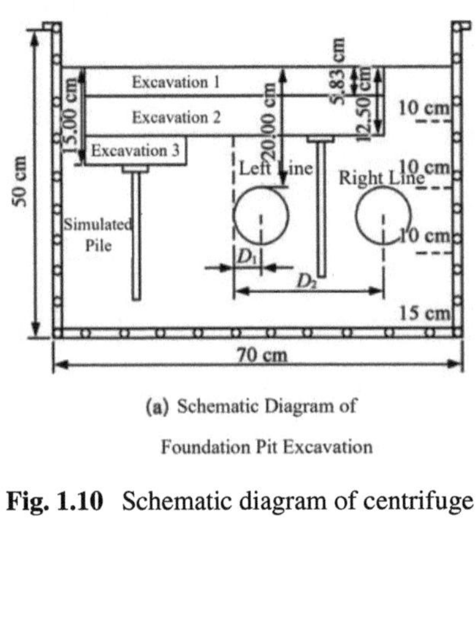

(a) Schematic Diagram of Foundation Pit Excavation

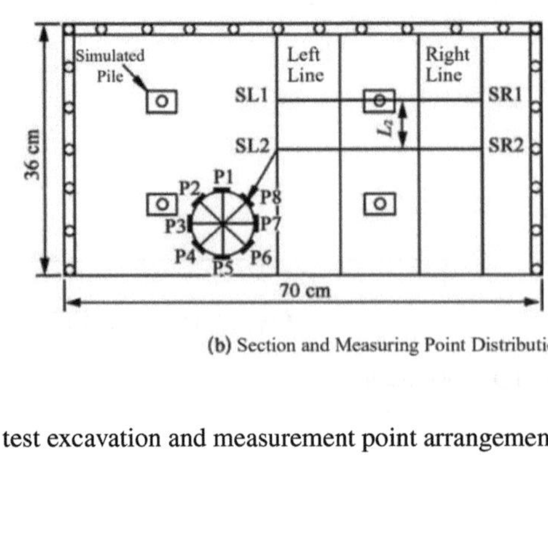

(b) Section and Measuring Point Distribution

Fig. 1.10 Schematic diagram of centrifuge test excavation and measurement point arrangement

Given the tunnel diseases, such as excessive convergence and joint damage, caused by loading/unloading, many scholars have also carried out experiments on tunnel reinforcement.

Bi et al. (2014) targeted the straight-jointed tunnels subjected to upper surcharge, conducting steel-plate reinforcement tests, and compared the mechanical characteristics between whole-ring and half-ring reinforcement. The results show that steel-plate reinforcement can improve the stiffness and bearing capacity of shield tunnels, and the effect of whole-ring reinforcement is close to that of half-ring reinforcement. The damage of both reinforcements is caused by localized adhesion failure of the steel rings. Liu et al. (2013) used full-scale tests to investigate the deformation properties and damage of shield tunnels treated by steel-plate reinforcement. The results show that the strength and stiffness of the shield tunnel increase dramatically before the adhesion between the steel ring and the lining concrete fails; the stiffness decreases rapidly after the failure, and the area around the capping block and the neighboring block is the weak area.

Liu et al. (2017) established a detailed finite element model of steel-plate reinforced shield tunnels based on full-scale tests. They increased the bearing capacity by 59% by conducting bonded steel plate reinforcement on a damaged stagger-jointed tunnel. They pointed out that the adhesion failure of epoxy resin between the bonded steel plate and the lining concrete causes brittle damage, which must be avoided.

Liu et al. (2015) used composite cavity, a new type of reinforcement material with lightweight and fast reinforcement, to reinforce the longitudinal joints in shield tunnels, and investigated the damage mechanism and reinforcement effect of the composite cavity for joint reinforcement. The results show that the reinforcement effect under negative bending moments is more significant than that under positive bending moments, and they also pointed out that sufficient attention should be paid to the bonding surface as a structural weak point. Ren et al. (2019) investigated the reinforcement effect of corrugated steel plate on shield tunnel lining segments based on full-scale tests, and found that corrugated steel reinforcement can increase the stiffness of the segments by 10–36% and the bearing capacity by 163–201%.

1.3 Research Content

Both large-scale surface loading and unloading will cause changes in the stress and displacement fields of soils within a certain range, breaking the original equilibrium of the shield tunnel. Once the additional stress field or displacement field caused by loading/unloading exceeds permissible limits, the tunnel may experience excessive transverse convergence and longitudinal settlement. When the deformation reaches a certain level, it can damage the tunnel structure and affect its normal operation. Therefore, the following aspects are introduced.

(1) Establish an analysis method for longitudinal deformation of subway tunnel under large-area loading based on the elastic theory.

For homogeneous soil, the analytical expressions of stress and displacement fields under surface loading can be obtained through application of the Boussinesq solution for vertical loads and double integration of surface surcharge effects, with corresponding theoretical analysis methods and solutions being derived accordingly. For layered soils, numerical solutions of stresses and displacements under surcharge are determined based on elastic semi-infinite space theory, while the longitudinal deformation response under surface surcharge is established through implementation of a two-stage method incorporating the Pasternak foundation model.

(2) Study the transverse deformation response under loading above the tunnel based on numerical models and laboratory tests.

A numerical model for transverse deformation response under large-area loading is established using ABAQUS finite element software, with particular emphasis placed on analyzing the influence of soil properties on transverse deformation. Laboratory model tests are conducted to investigate deformation responses and distribution characteristics of additional stresses around tunnels under eccentric loading conditions.

(3) Perform numerical analysis of the deformation behavior of shield tunnel under overhead unloading.

This book studies the effect of foundation pit excavation and unloading on tunnel deformation. Based on the pit excavation size, soil reinforcement effect, and the impact of anti-uplift piles and other factors, it analyzes the time–-space effects of foundation pit excavation on tunnel deformation through finite element numerical analysis. Combined with the engineering control measures of the neighboring subway tunnels, it summarizes the patterns of tunnel deformation triggered by overhead foundation pit excavation.

(4) Analyze on-site engineering cases.

Through long-term monitoring and analysis of typical loading and unloading cases, quantitative calculations of deformation amounts are performed and loading characteristics are determined. Research findings indicate that stratum heave can be effectively controlled within 1‰ H (where H represents the thickness of the unloaded soil layer) through implementation of time-limited unloading above the tunnel, foundation reinforcement measures, and installation of pile foundations to establish a comprehensive structural system for heave control.

References

Ai ZY, Wu C (2008) Transfer matrix solution for layered foundation in three-dimensional Cartesian coordinate system. J Chongqing Jianzhu Univ 30(02):43–46

Bahar L (1972) Transfer matrix approach to layered systems. Eng Mech Div ASCE 98(05):1159–1172

Bi XL, Liu X, Wang XZ et al (2014) Full-scale experimental study on ultimate bearing capacity of shield tunnel structures reinforced with inner steel circles. Chin Civil Eng J 47(11):128–137

Booker J, Small A (1982) Finite layer analysis of consolidation. Int J Numer Anal Methods Geomech 6(02):151–171

Bufler H (1971) Theory of elasticity of a multilayered medium. J Elast 1(02):125–143

Byun G, Kim D, Lee S (2006) Behavior of the ground in rectangularly crossed area due to tunnel excavation under the existing tunnel. Tunn Undergr Space Technol 21(3–4):361

Chen RP, Meng FY, Li ZC et al (2016) Excessive displacement and protection measures of metro tunnel adjacent to deep foundation pit. J Zhejiang Univ (Eng Sci) 50(05):856–863

Cheng H, Chen R, Wu H et al (2020) A simplified method for estimating the longitudinal and circumferential behaviors of the shield-driven tunnel adjacent to a braced excavation. Comput Geotechn 123:103595

Dai HW, Chen RP, Chen YM (2006) Analysis of new surface construction loads on longitudinal deformation of adjacent metro tunnels. Chin J Geotechn Eng 28(03):312–316

Fan YY, Guo XH, Deng ZJ et al (2014) Impact analysis of overlying loading/unloading on shield tunnel deformation. Constr Technol 43(07):107–109

Gao JJ (2017) Mechanical deformation characteristics and simplified algorithm of tunnels crossing intersection nodes. PhD dissertation, Shanghai Jiao Tong University, Shanghai

Gao JJ, Huang B, Zhang W et al (2018) Calculation method for vertical displacement of intersecting tunnels under surface surcharge. Tunn Constr (Chin Engl) 38(05):818–823

Ge SP, Xie DW, Ding WQ et al (2013) Simplified numerical simulation method for shield segment joints. Chin J Geotechn Eng 35(09):1600–1605

Guo J (2016) Efficiency study on segment lifting by grouting in shield tunnels. Railw Constr Technol 09:18–20

Huang X, Huang HW, Zhang DM (2012) Longitudinal deformation of underlying existing shield tunnels induced by excavation unloading. Chin J Geotechn Eng 34(07):1241–1249

Huang DW, Zhou SH, Feng QS et al (2018) Design of model test for interaction between shield tunnel and surrounding strata. J China Railw Soc 40(06):127–135

Hwang R, Chen B, Wu T (2015) Damage to a metro tunnel due to adjacent excavation. Springer India, New Delhi, pp 369–377

Jiang QY, Guan PF, Ye R (1999) Longitudinal deformation analysis of shield tunnels in soft soil. Undergr Eng Tunn 4:2–6

Kang C, Mei GX, Liang RZ et al (2018) Longitudinal deformation analysis of existing shield tunnel induced by temporary surface surcharge. Rock Soil Mech 39(12):4605–4616

Li ZF (2017) Study on the influence of deep excavation on metro structures. PhD dissertation, Chang'an University, Xi'an

Li CL, Wang GQ, Zhao KJ, Zhu CF (2011) Longitudinal mechanical behavior of shield tunnels under surface loading. J Jilin Univ (Eng Technol Ed) 41:180–184

Liang R, Xia T, Huang M et al (2017) Simplified analytical method for evaluating the effects of adjacent excavation on shield tunnel considering the shearing effect. Comput Geotechn 81(01):167–187

Liang FY, Yuan Q, Li JP, Zhang SX (2020) Influence of soil stratification characteristics on longitudinal deformation of metro tunnels under surcharge loading. Chin J Geotechn Eng 42(01):63–71

Liang F, Yuan Q, Song Z et al (2021a) Longitudinal responses of shield tunnel subjected to surcharge considering dislocation. Proc. Inst. Civ. Eng. Geotechn. Eng. 174(4):342–354

Liang FY, Fang YQ, Yuan Q et al (2021b) Experimental study on influence of local surcharge on transverse deformation of tunnels in soft and hard strata. J Tongji Univ (Nat Sci) 49(03):322–331

Lin YG, Liao SM, Liu GB (2000) Investigation on influencing factors of longitudinal deformation in subway tunnels. Undergr Space 20(04):264–267

Lin P, Xia HY, Zheng YW (2014) Analysis on treatment effect of tunnel deformation accident caused by foundation pit excavation. West-China Explor Eng 26(11):159–163

Liu X, Tang M, Lu L et al (2013) Experimental study on bearing capacity of shield tunnel structure reinforced with inner steel circles: complete ring reinforcement method. Chin J Rock Mech Eng 32(11):2300–2306

Liu X, Zhang CG, Zhang Y et al (2015) Experimental study on longitudinal joints of shield tunnel reinforced with composite cavity. J Railw Sci Eng 12(02):376–383

Liu TJ, Huang HH, Xu R et al (2017) Bearing performance of metro shield tunnel reinforced by bonded steel plates. China J Highw Transp 30(08):91–99

Loganathan N, Poulos H (1998) Analytical prediction for tunneling-induced ground movements in clays. J Geotechn Geoenviron Eng 124(9):846–856

Lu DY, Xu GW, Wang SM (2017) Effects of loading/unloading on material damage and structural characteristics of shield tunnels. J Southwest Jiaotong Univ 52(06):1104–1112

Peng YC, Ding WQ, Zhu HH et al (2013) Shell-joint model for shield tunnel lining structure. Chin J Geotechn Eng 35(10):1823–1829

Qi KJ, Wang XD, Jiang G et al (2005) Analysis of deep excavation adjacent to metro tunnels. Chin J Rock Mech Eng 24(S2):5485–5489

Qiu J, Liu H, Lai J et al (2018) Investigating the long-term settlement of a tunnel built over improved loessial foundation soil using jet grouting technique. J Perform Constr Facil 32(05):4018066

Qu JL, Pan R, Tang RD (2016) Study on tunnel longitudinal deformation based on double-sided elastic foundation beam. J Water Resour Water Eng 125(01):190–194

Ren TY, Liu SY, Liu X (2019) Experimental study on bending performance of shield tunnel lining segments reinforced with corrugated steel plates. Tunn Constr (Chin Engl) 39(02):317–323

Sang YL, Liu XZ, Zhang Q (2020) Analysis and application of joint stiffness in segmental rings of metro shield tunnels based on bolt-convex-concave Tenon connection. Tunn Constr (Chin Engl) 40(01):19–27

Shao H, Huang HW, Zhang DM et al (2016) Remediation of large deformation in soft soil subway shield tunnels induced by sudden surcharge. Chin J Geotechn Eng 38(06):1036–1043

Su ZX, He C (2015) Numerical analysis of additional internal forces in longitudinal deformation of shield tunnel using shell-spring-contact model. Mod Tunn Technol 52(06):70–76

Sun WB (2018) Finite element analysis of shield tunnel deformation under surface loading. Gansu Sci Technol 34(16):75–76

Verruijt A, Booker J (1996) Surface settlements due to deformation of a tunnel in an elastic half plane. Géotechnique 46(4):753–756

Wang K (1982) Mechanical calculation of N-layer elastic continuous system under circular uniform vertical load. Chin Civil Eng J 15(02):65–76

Wang T, Li H, Xu RQ (2005) Deformation analysis of shield tunnels induced by large-area surface loading/unloading. Mod Transp Technol 5(3):29–31

Wei G, Yu GH, Hong WQ (2018) Calculation of shear dislocation deformation of underlying shield tunnel induced by surface surcharge. J Cent South Univ (Sci Technol) 49(07):1775–1783

Wei G, Zhang XH, Lin XB et al (2020) Study on transverse mechanical behavior changes of adjacent shield tunnel induced by foundation pit excavation. Rock Soil Mech 41(02):635–644

Wu Q, Du SJ (2014) Experimental study on the influence of surface surcharge on existing shield tunnel structures. Chin J Undergr Space Eng 10(01):57–66

Wu SY, Yang XP, Liu TJ (2012) Influence of double-side deep excavation on deformation of adjacent metro tunnel. Chin J Rock Mech Eng 31(S1):3452–3458

Ye YD (2007) Structural deformation and health diagnosis method of operational shield tunnels in soft soil areas. PhD dissertation, Tongji University, Shanghai

Ying HW, Cheng K, Yu JL et al (2021) Displacement prediction of adjacent shield tunnel induced by excavation considering continuous ground deformation. J Zhejiang Univ (Eng Sci) 55(02):318–329

Zhang ZG, Huang MS, Wang WD (2009) Analysis of influence of new tunnel excavation on adjacent existing tunnel in layered soils. Chin J Geotechn Eng 31(4):600–608

Zhang ZG, Zhang MX, Wang WD (2011) Two-stage analysis method for influence of foundation pit excavation on adjacent metro tunnels. Rock Soil Mech 32(07):2085–2092

Zhang DM, Zou WB, Yan JY (2014) Lateral grouting control mechanism for large transverse deformation of shield tunnels in soft soil. Chin J Geotechn Eng 36(12):2203–2212

Zhang MG, Zhou SH, Huang DW et al (2016) Analysis of the influence of surface overload on metro shield tunnels. Rock Soil Mech 37(08):2271–2278

Zhang YW, Xie YL, Weng MS (2018) Centrifuge test on influence of asymmetric excavation on underlying metro tunnel. Rock Soil Mech 39(07):2555–2562

Zhang C, Zhu J, Huang M et al (2019a) Winkler load-transfer analysis for pipelines subjected to surface load. Comput Geotechn 111(07):147–156

Zhang D, Huang Z, Li Z et al (2019b) Analytical solution for the response of an existing tunnel to a new tunnel excavation underneath. Comput Geotech 108(04):197–211

Zhang D, Liu Z, Wang R et al (2019c) Influence of grouting on rehabilitation of an over-deformed operating shield tunnel lining in soft clay. Acta Geotechn 14(04):1227–1247

Zhang D, Zhai W, Huang H et al (2019d) Robust retrofitting design for rehabilitation of segmental tunnel linings: using the example of steel plates. Tunn Undergr Space Technol 83:231–242

Zhang DM, Liu J, Li BJ et al (2020) Shear resistance characteristics of inclined bolt ring joints in large-diameter shield tunnels. China J Highw Transp 33(12):142–153

Zhao H, Liu X, Bao Y et al (2016) Simplified nonlinear simulation of shield tunnel lining reinforced by epoxy bonded steel plates. Tunn Undergr Space Technol 51(01):362–371

Zheng G, Du YM, Diao Y et al (2016) Influence zone of deep excavation on adjacent existing tunnels. Chin J Geotechn Eng 38(04):599–612

Zhong Y, Zhang YS (2003) Exact stiffness matrix method for solving axisymmetric problems of multi-layered elastic half-space. Chin Q Mech 24(3):395–400

Zhong Y, Wang ZR, Guo DZ (1992) Transfer matrix method for solving axisymmetric problems of multi-layered elastic half-space. Chin Civil Eng J 25(6):37–42

Zhong Y, Wang ZR, Guo DZ (1995) Transfer matrix method for solving non-axisymmetric problems of multi-layered elastic half-space. Chin Civil Eng J 28(1):66–72

Zhou SY (2015) Numerical study on bearing performance of longitudinal joints in metro shield tunnels reinforced with steel plates. Master's thesis, South China University of Technology, Guangzhou

Chapter 2
Theoretical Analysis of Tunnel Longitudinal Deformation Under Surcharge Loading

2.1 Introduction

The additional stress caused by surface loading breaks the original stress equilibrium of the subway tunnel, triggering stress redistribution in the strata and tunnel, thus leading to longitudinal and transverse tunnel deformation. A number of engineering cases show that when the internal forces and deformation caused by surface loading exceed the allowable limits, the subway operation faces safety risks. Therefore, it is important to study the longitudinal deformation response of tunnels under surface loading.

Currently, scholars have theoretically studied tunnel longitudinal deformation. Most of them focus on the mechanical performance response of circumferential joints and on the deformation response of tunnels under adjacent unloading. However, their study on longitudinal deformation response under surface loading is incomplete due to insufficient quantitative assessment of the effects of soil stratification, surface loading, and other factors on tunnel deformation.

Based on the elastic semi-infinite space theory, this chapter derives the stress and displacement solutions for layered soils under surface loading; It introduces a theoretical calculation model of tunnel longitudinal deformation by using the equivalent stiffness model and the finite difference method; This chapter also analyzes the influence of various parameters, such as tunnel burial depth, loading magnitude and range, and soil properties on deformation, and explores and the patterns of tunnel longitudinal deformation under surface loading. The research will provide references for future engineering activities.

R. Wang et al., *Impact Analysis and Control Strategies of Overlying Loading/Unloading Effects on Soft Soil Shield Tunnels*, Key Technologies for Tunnel Construction under Complex Geological and Environmental Conditions,
https://doi.org/10.1007/978-981-95-1462-5_2

2.2 Theoretical Analytical Model for Longitudinal Deformation of Tunnels

2.2.1 *Foundation Models*

Given the analysis of soil-structure interaction, the Winkler (Huang et al. 2009; Li et al. 2016), Pasternak (Peng et al. 2020), and Kerr foundation models (Zhang et al. 2019a; Kerr 1985) along with the continuous elastic model (Klar et al. 2005) are commonly used for simulating the soft soil foundation. The Winkler model simplifies the soil as a series of independent soil springs, excluding the influence of shear force between the soil elements. This model, however, tends to underestimate the structural deformation of tunnels under external loading (Liang et al. 2017). By contrast, the Kerr model involves three undetermined parameters, but it is rather challenging to accurately determine these parameters. The continuous elastic model can reflect the diffusion characteristics of soil stress and provide relatively precise results. However, its calculation process is too complex, making it unsuitable for engineering applications. Compared to the three models aforementioned, the Pasternak model, a two-parameter foundation model, more effectively demonstrates the soil-tunnel interaction as it integrates the constraints between independent springs in the Winkler model to reflect the continuity of the soil. Based on extensive research on its relevant parameters, this book adopts the Pasternak foundation model to simulate the tunnel-soil interaction.

Under external loading, the relationship between the foundation reaction p and the soil displacement $w(x)$ in the Pasternak foundation model can be expressed as

$$P = k_s w(x) - G_s \frac{\partial^2 w(x)}{\partial x^2}, \tag{2.1}$$

where, k_S represents the foundation reaction coefficient; G_S is the foundation shear stiffness.

Based on Biot's research, Vesic (1961) analyzed the deformation characteristics of a beam on the ground under external loading and obtained

$$k_{\text{vesic}} = \frac{0.65E_s}{D_t(1 - v_s^2)} \sqrt[12]{\frac{E_s D_t^4}{EI}}. \tag{2.2}$$

This empirical formula for the foundation reaction coefficient has been widely used by scholars for analyzing the soil-structure interaction. In Eq. (2.2), E_S is the elastic modulus of soils; V_S is the soil Poisson's ratio; D_t is the width of the foundation beam; EI is the bending stiffness of the beam.

However, for beams at a certain depth, the Vesic empirical equation neglects the soils' constraint on beams, underestimating the foundation reaction coefficient (Yu et al. 2012, 2013) of soils to a certain degree. Ultimately, the structural deformation is overestimated. Based on this, Liang et al. (2018) and Zhang et al. (2019b) each introduced empirical formula for the foundation reaction coefficient that accounts for depth correction. Liang et al. (2021) applied both empirical formulas to calculate the deformation response of tunnels under surcharge. The results showed that although their expressions differ, the calculated results are essentially consistent. Therefore, to reflect the tunnel's burial depth, this book adopts the empirical formulas proposed by Zhang et al. (2019b) to determine the foundation reaction coefficient k_s:

$$k_s = \frac{4 - 5\nu_s}{\eta_1} \frac{0.65E_s}{D_t(1 - \nu_s^2)} \sqrt[12]{\frac{E_s D_t^4}{EI}}, \tag{2.3a}$$

$$\eta 1 = \begin{cases} 4 - 5\nu_s & z_0/D_t \leq 0.5 \\ 1 + D_t/z_0 & z_0/D_t > 0.5 \end{cases}. \tag{2.3b}$$

Here, η_1 is the depth correction factor for the foundation beam; z_O is the burial depth.

In practical engineering cases, the ratio of burial depth to tunnel diameter in Shanghai typically exceeds 0.5. Therefore, it is necessary to consider corrections to the foundation reaction coefficient.

Regarding the value of the shear stiffness of the soil G_P, many scholars have conducted extensive research (Kerr 1985; Tanahashi 2004; Liang et al. 2013). This book adopts the findings from Tanahashi (2004):

$$G_p = \frac{E_s H}{6(1 + \nu_s)}. \tag{2.4}$$

In this equation, H is the thickness of the effective compressible layer of soil. Xu (2005) used the Mindlin solution to conduct a detailed study on how the additional stress in soils under external loading varied with the burial depth. The results showed that the additional stress at a depth of 2.5 times the foundation beam width decreased to 20% of the external loading. In practical engineering, the elastic modulus of soils below 2.5 times the beam width is generally significantly higher than that of the soil directly beneath the foundation. Therefore, it is recommended that the depth of the effective compressible layer should be taken as $H = 2.5D_t$.

2.2.2 Tunnel Models

There are two commonly used beam models: Euler–Bernoulli and Timoshenko. The biggest difference between them is whether the shear deformation of the beam cross-section is considered, and the former is a special form of the latter, as shown in Fig. 2.1.

Urban subway tunnels are typically constructed from precast segments which are connected to each other with bolts to form a cohesive structure. Generally speaking, the presence of segment joints weakens the overall stiffness of tunnel structures. Extensive monitoring data from the fieldwork indicates that the deformation along the tunnel axis can be attributed to two parts (Fig. 2.2):

1. Bending deformation caused by bending, which leads to the opening of the segments.
2. Segment misalignment triggered by shear stress which is induced by differential settlement (Wang 2009; Shen et al. 2014; Wu et al. 2015). Dai et al. (2006) and Li et al. (2015) used the Euler–Bernoulli beam to calculate the longitudinal deformation of tunnels and pipelines. However, since the Euler–Bernoulli beam model only accounts for bending deformation and neglects the amount of segment misalignment, this book adopts the Timoshenko beam model to simulate shield tunnels.

To study the reduction in longitudinal stiffness of tunnels caused by segment joints, scholars have conducted extensive research (Ye et al. 2011; Shiba et al. 1988). Among these studies, the stiffness equivalence method proposed by Shiba, a Japanese scholar, is widely used in engineering practice for its clear concept and convenient

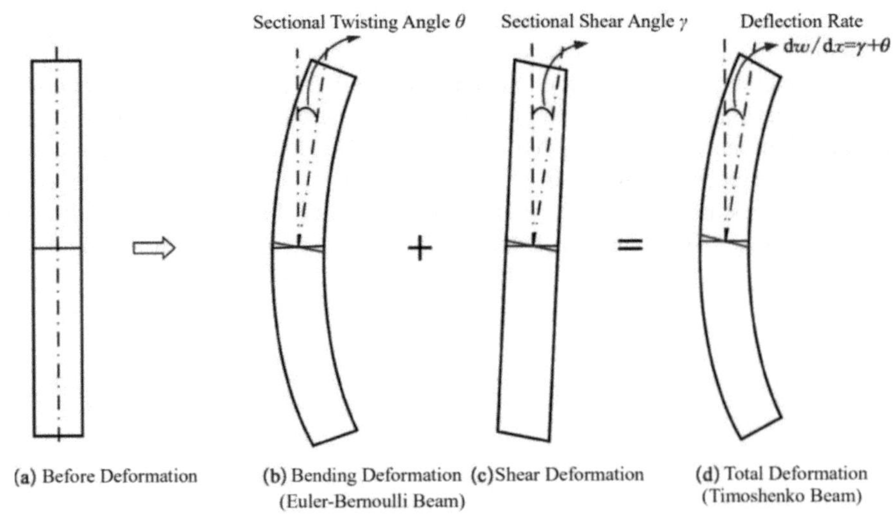

Fig. 2.1 Deformation patterns of beams

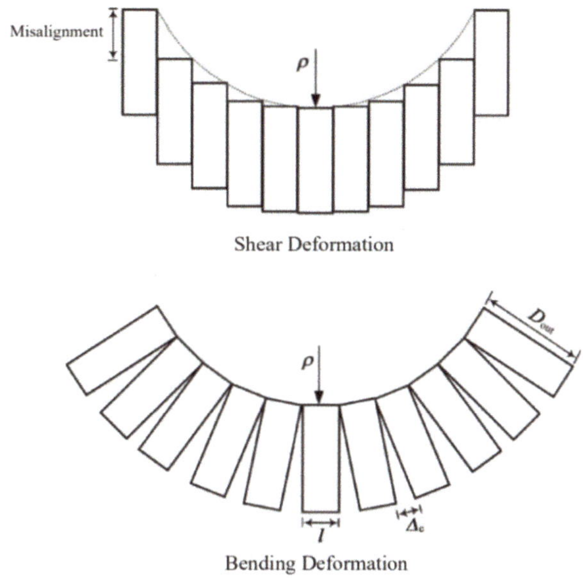

Fig. 2.2 Deformation patterns of tunnels

calculation, as shown in Fig. 2.3. The longitudinal stiffness equivalence of tunnels includes both bending and shear stiffness equivalence. The derivation process is as follows (Xu 2005; Shiba et al. 1988).

(1) Bending stiffness equivalence

To determine the longitudinal bending stiffness of shield tunnels, Shiba et al. (1988) assumed:

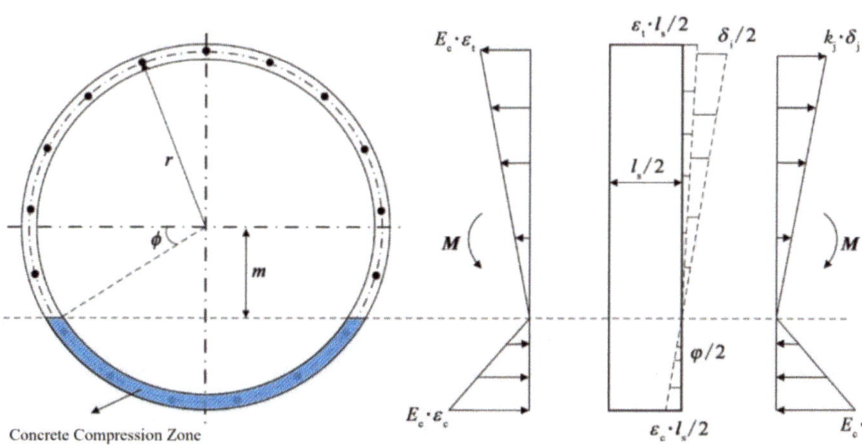

Fig. 2.3 Stress–strain diagram of segment ring

① The deformation at each point of the tunnel cross-section is proportional to its distance from the neutral axis, which is a plane section assumption;
② Under bending moment loads, the tensile force on the tension side of the segment ring is entirely borne by bolts, while the compressive force on the compression side is borne by segments;
③ From the interest of safety, the influence range of the joint is determined to be $l_S/2$, and the effect of axial forces on the bending stiffness of the joint is ignored;
④ The unevenly distributed longitudinal bolts are equivalent to a continuously distributed pattern along the circumferential direction. The average linear stiffness of bolts is

$$kj = \frac{n_b E_b A_b}{l_b 2\pi r},$$

where, n_b is the number of longitudinal bolts at the segment joint; E_b is the elastic modulus of the bolts; l_b is the length of the bolts; A_b is the cross-sectional area of the bolts.

According to the deformation compatibility conditions, Eqs. (2.5) and (2.6) can be obtained:

$$\frac{r+m}{D_t/2+m} \cdot \varepsilon_t \cdot \frac{l_s}{2} + \frac{\delta_j}{2} = (r+m)\frac{\varphi}{2}, \tag{2.5}$$

$$\varepsilon_c \cdot \frac{l_s}{2} = \left(\frac{D_t}{2} - m\right)\frac{\varphi}{2}. \tag{2.6}$$

According to the axial force equilibrium conditions, Eqs. (2.7) and (2.8) can be obtained:

$$2\frac{E_c\varepsilon_c}{D_t/2-m}\int_0^{\pi/2-\phi}(r\cos\alpha - m)rt\mathrm{d}\alpha = 2\frac{E_c\varepsilon_t}{D_t/2+m}\int_0^{\pi/2+\phi}(r\cos\alpha + m)rt\mathrm{d}\alpha, \tag{2.7}$$

$$2\frac{E_c\varepsilon_c}{D_t/2-m}\int_0^{\pi/2-\phi}(r\cos\alpha - m)rt\mathrm{d}\alpha = 2\frac{k_j\delta_j}{r+m}\int_0^{\pi/2+\phi}(r\cos\alpha + m)r\mathrm{d}\alpha. \tag{2.8}$$

In the equations, E_c is the elastic modulus of the segment; ε_c is the compressive strain of the segment; ε_t is the tensile strain of the segment; $m = r\sin\phi$ is the height of the neutral axis; ϕ is the angle corresponding to the neutral axis; φ is the rotation angle at the segment joint; l_s is the length of the segment; δ_j is the tensile deformation of the bolt; t is the thickness of the segment; r is the distance from the bolt to the tunnel axis.

According to the bending moment equilibrium conditions, Eq. (2.9) is as follows:

$$2\frac{E_c\varepsilon_c}{D_t/2-m}\int_0^{\pi/2-\phi}(r\cos\alpha-m)2rt\mathrm{d}\alpha+\frac{E_c\varepsilon_t}{D_t/2+m}\int_0^{\pi/2+\phi}(r\cos\alpha+m)2rt\mathrm{d}\alpha=M. \tag{2.9}$$

Given the deformation compatibility and the axial force equilibrium conditions, the position of the neutral axis can be determined as

$$\phi+\cot\phi=\pi\left(\frac{1}{2}+\frac{k_j l_s}{E_c t}\right). \tag{2.10}$$

Combing the deformation compatibility and the moment equilibrium conditions gives

$$\varphi=\frac{l_s}{E_c I_c}\cdot\frac{\cos\phi+(\phi+\pi/2)\sin\phi}{\cos^3\phi}M, \tag{2.11}$$

where, $I_c=\pi(D_t^4-D_i^4)/64$, which is the moment of inertia of the segment ring's cross-section about the tunnel axis.

Therefore, given the circumferential joints in the elastic state, the equivalent bending stiffness of the tunnel is as follows:

$$(EI)_{eq}=\frac{Ml_s}{\varphi}=E_c I_c\cdot\frac{\cos^3\phi}{\cos\phi+(\phi+\pi/2)\sin\phi}. \tag{2.12}$$

Note that the calculation model provided by Shiba has two limitations. First, the model fails to consider the influence of axial loading on the bending stiffness of segments, which may lead to an underestimation of the bending stiffness to some extent; Second, it neglects the reduction in bending stiffness caused by the yielding of segments and bolts. These two limitations regarding the longitudinal bending stiffness of the tunnel deserve further investigation.

The maximum opening width between segment rings caused by bending moment can be expressed as:

$$\Delta=\varphi(r+m)=\frac{Mr}{EI_{eq}}(1+\sin\phi)l_b. \tag{2.13}$$

(2) Equivalent Shear Stiffness

Wu et al. (2015) suggested that when the friction between segments at the circumferential joints was neglected, the total shear deformation u of segment rings under shear load Q is given by

$$u=u_j+u_s. \tag{2.14a}$$

in which u_j is the shear deformation of the bolts at the joint; u_s is the shear deformation of the segments.

From the definition of shear deformations, the following can be obtained

$$u = l_s \tan\left(\frac{Q}{(GA)_{eq}}\right), \tag{2.14b}$$

$$u_j = l_b \tan\left(\frac{Q}{n_b \kappa_b G_b A_b}\right), \tag{2.14c}$$

$$u_j = (l_s - l_b)\tan\left(\frac{Q}{\kappa_c G_c A_c}\right). \tag{2.14d}$$

In the equations, κ_b and κ_c are the shear coefficients for bolts and segments, respectively; For a circular section, κ_b is 0.9; for an annular (ring-shaped) section, κ_c is 0.5. G_b and G_C are shear modulus for bolts and segments, respectively.

By combining Eqs. (2.14a–d), Wu et al. (2015) derived

$$(GA)_{eq} = l_s\left(\frac{l_b}{n_b \kappa_b G_b A_b} + \frac{l_s - l_b}{\kappa_c G_c A_c}\right)^{-1}. \tag{2.15}$$

The equation for equivalent shear stiffness more comprehensively reflects the influencing factors, consistent with engineering practice.

The amount of misalignment between segments caused by shear can be expressed as:

$$\delta = l_s \frac{Q}{G_{eq}}. \tag{2.16}$$

2.2.3 *Calculation Models for Tunnel Deformation*

To simplify calculations, it is assumed that the tunnel deformation has been stabilized before the application of surface loading, and that the tunnel remains in continuous contact with the surrounding soils. And the temporal variation of soil consolidation, creep, and load are not considered. Based on Fig. 2.4, two equilibrium equations can be obtained:

$$\left.\begin{array}{l} \text{Force Equilibrium Equation: } Q + pD_t\,\mathrm{d}x = qD_t\mathrm{d}x + Q + \mathrm{d}Q \\ \text{Moment Equilibrium Equation: } -M - Q\mathrm{d}x - pD_t\,\mathrm{d}x\mathrm{d}x/2 + qD_t\,\mathrm{d}x\mathrm{d}x/2 \\ \qquad +M + \mathrm{d}M = 0 \end{array}\right\}, \tag{2.17}$$

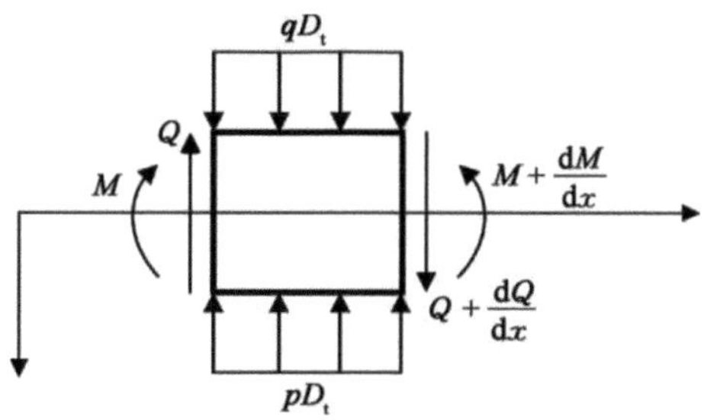

Fig. 2.4 Force analysis of tunnel differential unit

where, Q is the shear force on the tunnel; M is the bending moment on the tunnel; q is the additional stress caused by surcharge; D_t is the outer diameter of the tunnel; p is the foundation reaction force determined by Eq. (2.1); and $w(x)$ is the tunnel settlement caused by surcharge.

According to the Timoshenko beam theory, the relationship between internal forces and deformations can be expressed as:

$$M = -EI_{eq}\frac{d\theta}{dx}, Q = G_{eq}\left(\frac{dw}{dx} - \theta\right), \tag{2.18}$$

where, θ is the rotation angle of the neutral axis of the beam cross-section.

By solving Eqs. (2.17) and (2.18), the differential governing equation for the longitudinal displacement $w(x)$ of shield tunnels under longitudinal surface loading can be written as

$$\frac{d^4w(x)}{dx^4} + \alpha\frac{d^2w(x)}{dx^2} + \beta w(x) = \alpha 1\frac{d^2q(x)}{dx^2} + \beta 1 q(x), \tag{2.19}$$

where

$$\alpha = -\frac{k_s D_t EI_{eq} + D_t G_s G_{eq}}{(G_{eq} + D_t G_s)EI_{eq}};$$

$$\beta = \frac{G_{eq} k_s D_t}{(G_{eq} + D_t G_s)EI_{eq}};$$

$$\alpha 1 = -\frac{D_t}{G_{eq} + D_t G_s};$$

$$\beta 1 = \frac{D_t G_{eq}}{EI_{eq}(G_{eq} + D_t G_s)}.$$

Equation (2.19) is a fourth-order ordinary differential equation, so it is difficult to obtain an explicit analytical solution. Based on FEM, it can be transformed into a simple algebraic equation, and be solved programmatically to yield its numerical solution. First, the tunnel is discretized into $n + 5$ differential elements, with the

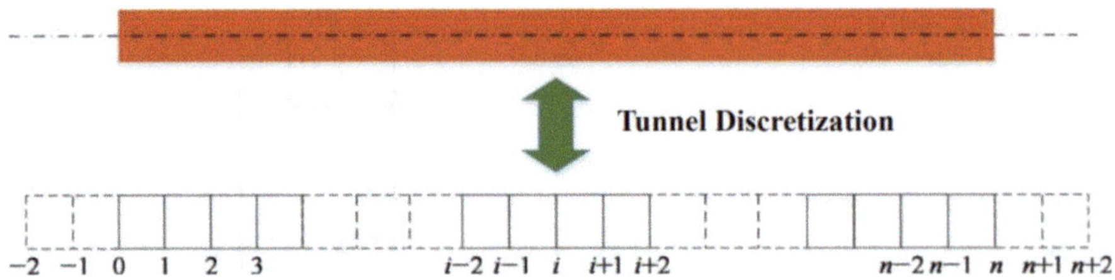

Fig. 2.5 Discretization analysis of tunnel

length of each unit set to l (see Fig. 2.5). The points labeled -2, -1, $n+1$, and $n+2$ are virtual nodes. It is assumed that both ends of the tunnel are free, meaning there is no bending moment or shear force acting at the ends. Therefore, for the nodes at 0 and n, two boundary conditions are satisfied: $Q_0 = Q_n = 0$; $M_0 = M_n = 0$. Based on the finite difference principle and the boundary conditions, Eq. (2.19) can be rewritten as

$$KW = Dq. \tag{2.20}$$

In this vector–matrix expression, the deflection $w(x)$ is an unknown variable, and

$$W = \begin{bmatrix} w_0 \; w_1 \; \ldots \; w_{n-1} \; w_n \end{bmatrix}^{\mathrm{T}};$$
$$S = \begin{bmatrix} q_0 \; q_1 \; \ldots \; q_{n-1} \; q_n \end{bmatrix}^{\mathrm{T}}.$$

$$K = \begin{bmatrix} \frac{2+l^4\beta}{l^4} & \frac{-4+2l^2\alpha}{l^4} & \frac{2}{l^4} & & & & & \\ \frac{-2}{l^4} & \frac{5-2l^2\alpha+l^4\beta}{l^4} & \frac{-4+l^2\alpha}{l^4} & \frac{1}{l^4} & & & & \\ \frac{1}{l^4} & \frac{-4+l^2\alpha}{l^4} & \frac{6}{l^4}-\frac{2\alpha}{l^2}+\beta & \frac{-4+l^2\alpha}{l^4} & \frac{1}{l^4} & & & \\ & \ddots & \ddots & \ddots & \ddots & \ddots & & \\ & & \ddots & \ddots & \ddots & \ddots & \ddots & \\ & & & \frac{1}{l^4} & \frac{-4+l^2\alpha}{l^4} & \frac{6}{l^4}-\frac{2\alpha}{l^2}+\beta & \frac{-4+l^2\alpha}{l^4} & \frac{1}{l^4} \\ & & & & \frac{1}{l^4} & \frac{-4+l^2\alpha}{l^4} & \frac{5-2l^2\alpha+l^4\beta}{l^4} & \frac{-2}{l^4} \\ & & & & & \frac{2}{l^4} & \frac{-4+2l^2\alpha}{l^4} & \frac{2+l^4\beta}{l^4} \end{bmatrix}_{(n+1)\times(n+1)}$$

$$
D = \begin{bmatrix}
\beta_1 & \frac{2\alpha_1}{l^2} & & & & & \\
0 & \beta_1 - \frac{2\alpha_1}{l^2} & \frac{\alpha_1}{l^2} & & & & \\
 & \frac{\alpha_1}{l^2} & \beta_1 - \frac{2\alpha_1}{l^2} & \frac{\alpha_1}{l^2} & & & \\
 & & \ddots & \ddots & \ddots & & \\
 & & & \frac{\alpha_1}{l^2} & \beta_1 - \frac{2\alpha_1}{l^2} & \frac{\alpha_1}{l^2} & \\
 & & & & \frac{\alpha}{l^2} & \beta_1 - \frac{2\alpha_1}{l^2} & 0 \\
 & & & & & \frac{2\alpha_1}{l^2} & \beta_1
\end{bmatrix}_{(n+1)\times(n+1)}
$$

By solving Eq. (2.20), the tunnel longitudinal deformation under surface loading can be obtained. And integrating Eqs. (2.13) and (2.16) gives the solutions of bending moments, shear forces, opening width, and misalignment (Liang et al. 2020).

2.2.4 Calculation of Additional Stress in Tunnels Under Surface Loading

Figure 2.6 demonstrates the deformation response of underlying operating tunnels under surface loading. The surface surcharge is assumed to be evenly distributed with a magnitude of p, covering an area of $L_1 \times L_2$ (where, L_2 aligns with the tunnel axis, and L_1 is perpendicular to the axis). The outer diameter of the tunnel is D_t, and the distance between the tunnel axis and the surface is $z_0 = 10$ m. Based on the analysis above, the Pasternak foundation model is used to simulate the soil, and the Timoshenko beam model is used for the tunnel (Liang et al. 2020). To determine the tunnel longitudinal deformation response under external loading, the following assumptions are made:

① Before the application of external loading, the tunnel is in equilibrium, and any existing deformation is neglected.
② The plastic deformation of soils is neglected, meaning that the soil surrounding the tunnel remains in an elastic state under external loading.
③ Throughout the deformation process, the tunnel remains in continuous contact with the surrounding soil.
④ The presence of the tunnel does not affect the free stress field in soils induced by surface surcharge.

(1) Homogeneous soil foundation

In homogeneous soil strata, the additional free stress field in the soil under surcharge is treated as a semi-infinite elastic space, so it can be calculated by using the Boussinesq solution:

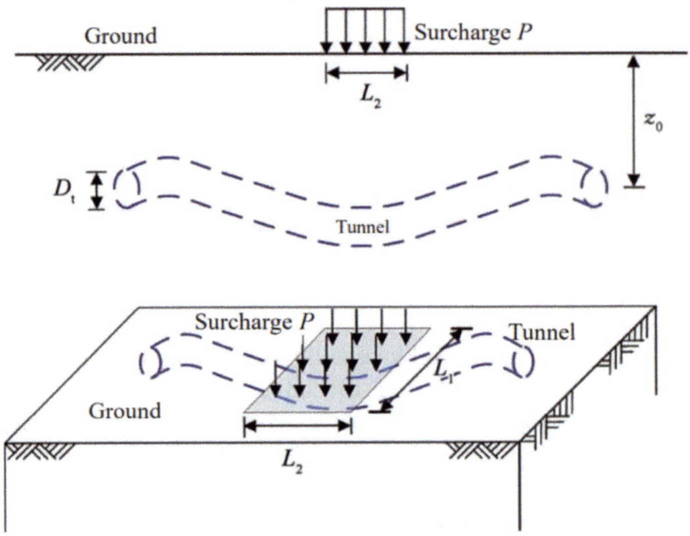

Fig. 2.6 Schematic of tunnel deformation under surface loading

$$q(x, y, z) = \int_{\frac{L_1}{2}}^{\frac{L_1}{2}} \int_{\frac{L_2}{2}}^{\frac{L_2}{2}} \frac{3z_0^3 p}{2\pi R^5} \mathrm{d}\eta \mathrm{d}\xi, \tag{2.21}$$

where, $R = \sqrt{(X - \xi)^2 + (Y - \eta)^2 + z_0^2}$. X and Y respectively represent the horizontal and vertical coordinates in the ξ–η coordinate system, as illustrated in Fig. 2.7.

The relationship between the x–y coordinate (tunnel coordinate system) and the ξ–η coordinate (load coordinate system) can be given by using a coordinate transformation matrix:

Fig. 2.7 Coordinate transformation

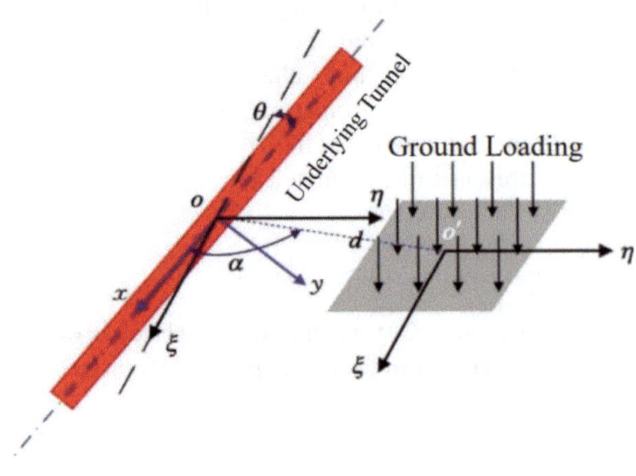

$$\begin{cases} X = x\cos\theta + y\sin\theta - d\cos\alpha \\ Y = -x\sin\theta + y\cos\theta - d\sin\alpha \end{cases}. \tag{2.22}$$

Here, α represents the angle between the line *oo'* and the *x*-axis, and θ is the angle between the *x*-axis and ξ-axis.

Since the Timoshenko beam model is used to simulate the tunnel, to determine the additional load acting on the tunnel under surcharge, the tunnel diameter shall be multiplied by Eq. (2.21).

(2) Layered soils

Based on the semi-infinite elastic body theory, some researchers (Ai and Wu 2008) have proposed solutions to determine the foundation stresses in layered soils. See the derivation process.

Under rectangular loading, the equilibrium equations for a soil micro-element are:

$$\frac{\partial\sigma_x}{\partial x} + \frac{\partial\tau_{xy}}{\partial y} + \frac{\partial\tau_{xz}}{\partial z} = 0 \tag{2.23a}$$

$$\frac{\partial\tau_{yx}}{\partial x} + \frac{\partial\sigma_y}{\partial y} + \frac{\partial\tau_{yz}}{\partial z} = 0 \tag{2.23b}$$

$$\frac{\partial\tau_{zx}}{\partial x} + \frac{\partial\tau_{zy}}{\partial y} + \frac{\partial\sigma_z}{\partial z} = 0 \tag{2.23c}$$

$$\sigma_{ij,i} = 0 \tag{2.23d}$$

It is assumed that soil particles are incompressible and soil deformation is linear elastic. Based on the generalized Hooke's Law, the following can be obtained

$$\sigma ij = 2G\left(\varepsilon_{ij} + \frac{\mu}{1-2\mu}\varepsilon_v\delta_{ij}\right), \tag{2.24a}$$

$$\varepsilon_{ij} = \frac{1}{E}\left[(1+\mu)\sigma_{ij} - \mu\sigma_{kk}\delta_{ij}\right]. \tag{2.24b}$$

The geometric equation is:

$$\varepsilon_{ij} = \frac{1}{2}\left(u_{i,j} + u_{j,i}\right) \tag{2.25}$$

Substituting Eqs. (2.24a, b) and (2.25) into Eqs. (2.23a–d) yields

$$\left.\begin{aligned}
\frac{\partial \sigma_z}{\partial z} &= -\frac{\partial \tau_{zx}}{\partial x} - \frac{\partial \tau_{yz}}{\partial y}\\
\frac{\partial \tau_{xz}}{\partial z} &= -\frac{E}{1-\mu^2}\frac{\partial^2 u_x}{\partial x^2} - \frac{E}{2(1-\mu)}\frac{\partial^2 u_y}{\partial x \partial y} - \frac{E}{2(1+\mu)}\frac{\partial^2 u_x}{\partial y^2} - \frac{\mu}{1-\mu}\frac{\partial \sigma_z}{\partial x}\\
\frac{\partial \tau_{yz}}{\partial z} &= -\frac{E}{1-\mu^2}\frac{\partial^2 u_y}{\partial y^2} - \frac{E}{2(1-\mu)}\frac{\partial^2 u_x}{\partial x \partial y} - \frac{E}{2(1+\mu)}\frac{\partial^2 u_y}{\partial x^2} - \frac{\mu}{1-\mu}\frac{\partial \sigma_z}{\partial y}\\
\frac{\partial u_x}{\partial z} &= \frac{2(1+\mu)}{E}\tau_{xz} - \frac{\partial u_z}{\partial x}\\
\frac{\partial u_y}{\partial z} &= \frac{2(1+\mu)}{E}\tau_{yz} - \frac{\partial u_z}{\partial y}\\
\frac{\partial u_z}{\partial z} &= \frac{(1+\mu)(1-2\mu)}{(1-\mu)E}\sigma_z - \frac{\mu}{1-\mu}\frac{\partial u_x}{\partial x} - \frac{\mu}{1-\mu}\frac{\partial u_y}{\partial y}
\end{aligned}\right\} \tag{2.26}$$

Transform the expression

$$\chi 1 = \frac{\partial \tau_{xz}}{\partial x} + \frac{\partial \tau_{yz}}{\partial y},$$

$$\Gamma 1 = \frac{\partial u_x}{\partial x} + \frac{\partial u_y}{\partial y},$$

$$\chi 2 = \frac{\partial \tau_{xz}}{\partial y} - \frac{\partial \tau_{yz}}{\partial x},$$

$$\Gamma 2 = \frac{\partial u_x}{\partial y} - \frac{\partial u_y}{\partial x}.$$

into Eq. (2.25), and obtain

$$\left.\begin{aligned}
\frac{\partial \chi_1}{\partial z} &= -\frac{E}{1-\mu^2}\Delta^2\Gamma_1 - \frac{\mu}{1-\mu}\Delta^2\sigma_z\\
\frac{\partial \sigma_z}{\partial z} &= -\chi_1\\
\frac{\partial \Gamma_1}{\partial z} &= \frac{2(1+\mu)}{E}\chi_1 - \Delta^2 u_z\\
\frac{\partial u_z}{\partial z} &= \frac{(1+\mu)(1-2\mu)}{(1-\mu)E}\sigma_z - \frac{\mu}{1-\mu}\Gamma_1
\end{aligned}\right\}, \tag{2.27a}$$

$$\left.\begin{aligned}
\frac{\partial \chi_2}{\partial z} &= -\frac{E}{2(1+\mu)}\Delta^2\Gamma_2\\
\frac{\partial \Gamma_2}{\partial z} &= \frac{2(1+\mu)}{E}\chi_2
\end{aligned}\right\}. \tag{2.27b}$$

The double Fourier transform and its inverse are as follows:

$$(\overline{\chi}_1, \overline{\Gamma}_1, \underline{\sigma}_z, \underline{u}_z, \overline{\chi}_2, \overline{\Gamma}_2) = \frac{1}{4\pi^2}\int_{-\infty}^{\infty}\int_{-\infty}^{\infty}(\chi_1, \Gamma_1, \sigma_z, u_z, \chi_2, \Gamma_2)e^{i(f_x x+f_y y)}\mathrm{d}x\mathrm{d}y, \tag{2.28a}$$

$$(\mathrm{X}_1, \Gamma_1, \sigma_z, u_z, \chi_2, \Gamma_2) = \int_{-\infty}^{\infty}\int_{-\infty}^{\infty}(\overline{\chi}_1, \overline{\Gamma}_1, \underline{\sigma}_z, \underline{u}_z, \overline{\chi}_2, \overline{\Gamma}_2)e^{-i(f_x x+f_y y)}\mathrm{d}fx\mathrm{d}fx. \tag{2.28b}$$

The Fourier transform of Eqs. (2.27a, b) can be written in a matrix form:

$$\frac{\mathrm{d}}{\mathrm{dz}}\begin{bmatrix}\Gamma_1\\ u_z\\ \chi_1\\ \sigma_z\end{bmatrix}=\begin{bmatrix}0 & f^2\,\frac{2(1+\mu)}{E} & 0 & \\ -\frac{\mu}{1-\mu} & 0 & 0 & \frac{(1+\mu)(1-2\mu)}{(1-\mu)E}\\ \frac{E}{1-\mu^2}f^2 & 0 & 0 & \frac{\mu}{1-\mu}f^2\\ 0 & 0 & -1 & 0\end{bmatrix}\begin{bmatrix}\Gamma_1\\ u_z\\ \chi_1\\ \sigma_z\end{bmatrix}, \tag{2.29a}$$

$$\frac{\mathrm{d}}{\mathrm{dz}}\begin{bmatrix}\Gamma_2\\ \chi_2\end{bmatrix}=\begin{bmatrix}0 & \frac{2(1+\mu)}{E}\\ \frac{E}{2(1+\mu)}f^2 & \end{bmatrix}\begin{bmatrix}\Gamma_2\\ \chi_2\end{bmatrix}. \tag{2.29b}$$

Equations (2.29a, b) can be simplified to give

$$\frac{\mathrm{d}\overline{U}_{\mathrm{I1}}}{\mathrm{d}z}=A_{\mathrm{I2}}\overline{U}_{\mathrm{I1}}, \tag{2.30a}$$

$$\frac{\mathrm{d}\overline{U}_{\mathrm{I2}}}{\mathrm{d}z}=A_{\mathrm{I2}}\overline{U}_{\mathrm{I2}}, \tag{2.30b}$$

where

$$\overline{U}_{\mathrm{I1}}=\left[\overline{\Gamma}_1\ \underline{u}_z\ \overline{\chi}_1\ \underline{\sigma}_z\right]^{\mathrm{T}},$$

$$\overline{U}_{\mathrm{I2}}=\left[\overline{\Gamma}_2\ \overline{\chi}_2\right]^{\mathrm{T}},$$

$$A_{\mathrm{I1}}=\begin{bmatrix}0 & f^2\,\frac{2(1+\mu)}{E} & 0 & \\ -\frac{\mu}{1-\mu} & 0 & 0 & \frac{(1+\mu)(1-2\mu)}{(1-\mu)E}\\ \frac{E}{1-\mu^2}f^2 & 0 & 0 & \frac{1}{1-\mu^2}f^2\\ 0 & 0 & -1 & 0\end{bmatrix}\begin{bmatrix}\Gamma_1\\ u_z\\ \chi_1\\ \sigma_z\end{bmatrix},$$

$$A_{\mathrm{I2}}=\begin{bmatrix}0 & \frac{2(1+\mu)}{E}\\ \frac{E}{2(1+\mu)}f^2 & \end{bmatrix}.$$

Based on the ordinary differential equation theory, the solutions to Eqs. (2.30a, b) can be written as

$$\overline{U}_{\mathrm{I1}}(f,z)=e^{\mathrm{A_{I1}}z}\overline{U}_{\mathrm{I1}}(f,0)=Q(f,z)\overline{U}_{\mathrm{I1}}(f,0), \tag{2.31a}$$

$$\overline{U}_{\mathrm{I2}}(f,z)=e^{\mathrm{A_{I2}}z}\overline{U}_{\mathrm{I2}}(f,0)=M(f,z)\overline{U}_{\mathrm{I2}}(f,0). \tag{2.31b}$$

Based on the Cayley-Hamilton theorem, matrices Q and M can be determined. The matrix form of Eqs. (2.30a, b) is

$$\begin{bmatrix} \overline{\Gamma}_1(f, z) \\ \underline{u}_z(f, z) \\ \overline{\chi}_1(f, z) \\ \underline{\sigma}_z(f, z) \end{bmatrix} = \begin{bmatrix} Q_{11} & Q_{12} & Q_{13} & Q_{14} \\ Q_{21} & Q_{22} & Q_{23} & Q_{24} \\ Q_{31} & Q_{32} & Q_{33} & Q_{34} \\ Q_{41} & Q_{42} & Q_{43} & Q_{44} \end{bmatrix} \begin{bmatrix} \overline{\Gamma}_1(f, 0) \\ \underline{u}_z(f, 0) \\ \overline{\chi}_1(f, 0) \\ \underline{\sigma}_z(f, 0) \end{bmatrix}, \tag{2.32a}$$

$$\begin{bmatrix} \overline{\Gamma}_2(f, z) \\ \overline{\chi}_2(f, z) \end{bmatrix} = \begin{bmatrix} s_{11} & s_{12} \\ s_{21} & s_{22} \end{bmatrix} \begin{bmatrix} \overline{\Gamma}_2(f, 0) \\ \overline{\chi}_2(f, 0) \end{bmatrix}. \tag{2.32b}$$

Transform Eqs. (2.32a, b), and obtain the stiffness matrix similar to that used in finite element analysis:

$$\begin{bmatrix} -\overline{\chi}_1(f, 0) \\ -\underline{\sigma}_z(f, 0) \\ \overline{\chi}_1(f, z) \\ \underline{\sigma}_z(f, z) \end{bmatrix} = \begin{bmatrix} K_1 & K_2 \\ K_3 & K_4 \end{bmatrix} \begin{bmatrix} \overline{\Gamma}_1(f, 0) \\ \underline{u}_z(f, 0) \\ \overline{\Gamma}_1(f, z) \\ \underline{u}_z(f, z) \end{bmatrix}, \tag{2.33a}$$

$$\begin{bmatrix} -\overline{\chi}_2(f, 0) \\ \overline{\chi}_2(f, z) \end{bmatrix} = \begin{bmatrix} \frac{s_{11}}{s_{12}} & -\frac{1}{s_{12}} \\ s_{21} - \frac{s_{22}s_{11}}{s_{12}} & \frac{s_{22}}{s_{12}} \end{bmatrix} \begin{bmatrix} \overline{\Gamma}_2(f, 0) \\ \overline{\Gamma}_2(f, z) \end{bmatrix}, \tag{2.33b}$$

where

$$K_1 = \begin{bmatrix} Q_{13} & Q_{14} \\ Q_{23} & Q_{24} \end{bmatrix}^{-1} \begin{bmatrix} Q_{11} & Q_{12} \\ Q_{21} & Q_{22} \end{bmatrix};$$

$$K_2 = -\begin{bmatrix} Q_{13} & Q_{14} \\ Q_{23} & Q_{24} \end{bmatrix}^{-1};$$

$$K_3 = \begin{bmatrix} Q_{31} & Q_{32} \\ Q_{41} & Q_{42} \end{bmatrix} - \begin{bmatrix} Q_{33} & Q_{34} \\ Q_{43} & Q_{44} \end{bmatrix} \begin{bmatrix} Q_{13} & Q_{14} \\ Q_{23} & Q_{24} \end{bmatrix}^{-1} \begin{bmatrix} Q_{11} & Q_{12} \\ Q_{21} & Q_{22} \end{bmatrix};$$

$$K_4 = \begin{bmatrix} Q_{33} & Q_{34} \\ Q_{43} & Q_{44} \end{bmatrix} \begin{bmatrix} Q_{13} & Q_{14} \\ Q_{23} & Q_{24} \end{bmatrix}^{-1}.$$

For the multilayer elastic space problem (Fig. 2.8), based on Eqs. (2.33a, b), the stress-displacement relationship for any given layer (the *j-th* layer) can be expressed as:

$$\begin{bmatrix} -\overline{\chi}_1(f, h_{j-1}) \\ -\underline{\sigma}_z(f, h_{j-1}) \\ \overline{\chi}_1(f, h_j) \\ \underline{\sigma}_z(f, h_j) \end{bmatrix} = K^j \begin{bmatrix} \overline{\Gamma}_1(f, h_{j-1}) \\ \underline{u}_z(f, h_{j-1}) \\ \overline{\Gamma}_1(f, h_j) \\ \underline{u}_z(f, h_j) \end{bmatrix}. \tag{2.34}$$

When solving such kinds of problems, all displacement and stress components approach zero if the bottom layer is an elastic half-space, within the elastic half-space

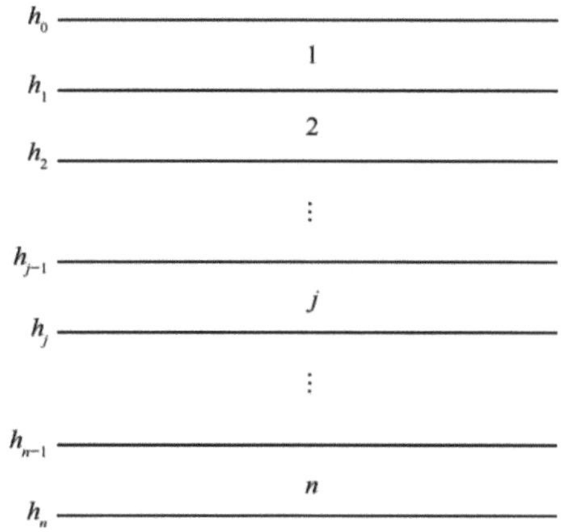

Fig. 2.8 Multilayer elastic space problem

as $z \to \infty$. Each element in Q contains terms e^{-kz} and e^{2kz}, which, however, fails to satisfy the boundary conditions of the semi-infinite space, so e^{2kz} is removed. Setting $z = 0$, the stress-displacement relationship at the top of the semi-infinite space can be obtained as

$$\begin{bmatrix} -\overline{\chi}_1(f, h_{n-1}) \\ -\underline{\sigma}_z(f, h_{n-1}) \end{bmatrix} = K^n \begin{bmatrix} \overline{\Gamma}_1(f, h_{n-1}) \\ \underline{u}_z(f, h_{n-1}) \end{bmatrix}. \tag{2.35}$$

Combing Eqs. (2.34) and (2.35) gives the stress-displacement relationship in layered soils:

$$\begin{bmatrix} -\overline{\chi}_1(f, 0) \\ -\underline{\sigma}_z(f, 0) \\ 0 \\ \cdot \\ \cdot \\ \cdot \\ 0 \\ 0 \end{bmatrix} = \begin{bmatrix} \boxed{K^1} & & & & & \\ & \boxed{K^2} & & & & \\ & & \cdot & & & \\ & & & \boxed{K^j} & & \\ & & & & \cdot & \\ & & & & \boxed{K^{n-1}} & \\ & & & & & \boxed{K^n} \end{bmatrix} \begin{bmatrix} \overline{\Gamma}_1(f, 0) \\ \underline{u}_z(f, 0) \\ \cdot \\ \overline{\Gamma}_1(f, h_{j-1}) \\ \underline{u}_z(f, h_{j-1}) \\ \cdot \\ \overline{\Gamma}_1(f, h_{n-1}) \\ \underline{u}_z(f, h_{n-1}) \end{bmatrix}. \tag{2.36}$$

By applying the inverse double Fourier transform to Eq. (2.36), the solution for each component within the soil layers in the real spatial domain under surface loading can be obtained. The detailed derivation process can be found in the research by Ai and Wu (2008).

Based on Eqs. (2.28a, b), using the inverse double Fourier transform and Gaussian integration method allows us to obtain the additional stress caused by surface loading in a layered foundation. The accuracy of the layered foundation stress calculation model and numerical transformation can be verified by comparing it with the Boussinesq solution. Assume that the surface load is 100 kN covering an area of 4 ×

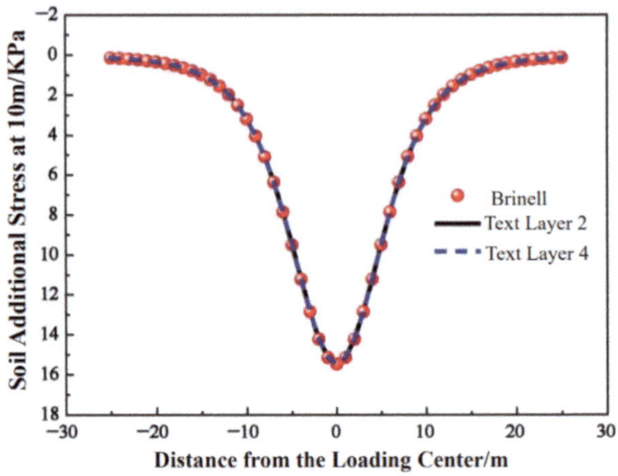

Fig. 2.9 Comparison of additional stress calculation results

10 m, and the foundation soil is an elastic semi-infinite body with an elastic modulus of 10 MPa and a Poisson's ratio of 0.33. The additional stress at a depth of 10 m is calculated using both the Boussinesq solution and the proposed solutions mentioned above (with soil layers set to two and four layers). The results are shown in Fig. 2.9. The comparison results show that the calculations in this book are consistent with the Boussinesq solution, validating the reliability of the calculation model of additional stress in layered soil presented in this book.

2.2.5 Model Correctness Verification

In a certain river channel in Shanghai with a depth of 3 m and a width of 24 m along the tunnel axis, the operating subway tunnel is buried at a depth of 8.1 m. The external diameter of the tunnel is ϕ6.2 m, and the internal diameter is ϕ5.5 m. The segments are made of C50 concrete with an elastic modulus of 34.5GPa. The equivalent bending stiffness is 7.8×10^7 kN·m^2, and the equivalent shear stiffness is 2.4×10^6 kN. The elastic modulus of the soil is 15 MPa with a Poisson's ratio of 0.3. To facilitate the construction of a nearby viaduct, the river channel is used as its storage yard, as shown in Fig. 2.10. Field monitoring data indicated that after backfilling the river channel, significant differential settlement occurred in the underlying tunnel, and water leakage was observed inside the tunnel (Fan et al. 2014).

The Boussinesq theory is used to solve the additional stress field at the tunnel axis caused by the river channel backfilling, while the Timoshenko beam + Pasternak foundation model is used to address the differential settlement induced by the surcharge. The calculation results are shown in Fig. 2.11. Comparing the theoretical values with the field measurements in Fig. 2.11, it can be seen that they are in

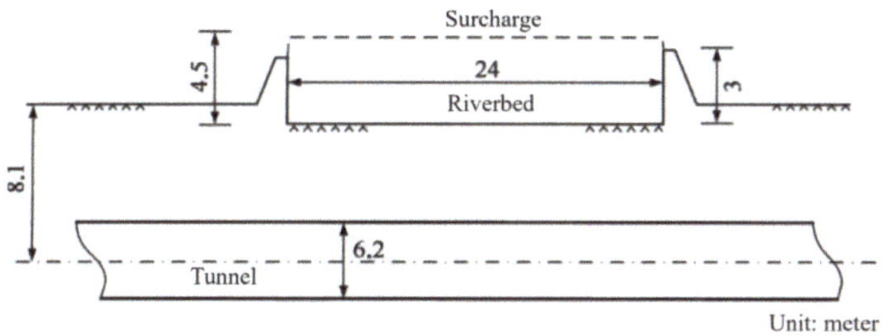

Fig. 2.10 Impact of backfilling on the tunnel

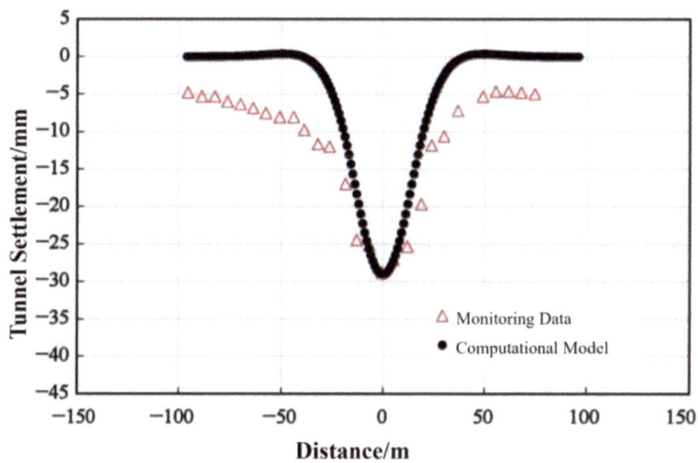

Fig. 2.11 Comparison of calculation results and field measurement data

good agreement, and the overall trend matches well. Therefore, the correctness and effectiveness of the calculation model in this book are validated.

2.3 Analysis of Parameters

The longitudinal deformation calculation models in Sect. 2.2 are used for parameter analysis of the effects of factors, including tunnel burial depth, surcharge magnitude, surcharge eccentricity, tunnel shear stiffness, and soil properties, on the longitudinal deformation of the underlying operating tunnel. The tunnel is modeled as a typical shield tunnel in Shanghai. The tunnel parameters are listed in Table 2.1. Based on the calculation methods in Sect. 2.2.2, the equivalent bending stiffness is determined to be 7.8×10^7 kN·m^2, and the equivalent shear stiffness to be 2.4×10^6 kN.

Table 2.1 Tunnel parameters

	D_t/mm	D_i/mm	t/mm	l_s/mm	E_c/MPa	D_b/mm	l_b/mm	E_b/MPa	n_b
Tunnel	6200	5500	350	1.2	3.45×10^4	30	400	2.06×10^5	17

Note D_t and D_i are the outer and inner diameters of the tunnel, respectively; t is the thickness of segments; ls is the ring width of segments; Ec and Eb are the elastic moduli of segments and bolts, respectively; Db and lb are the diameter and length of bolts, respectively

2.3.1 Tunnel Burial Depth

Most shield tunnels in Shanghai are laid at moderate depths. For the tunnel axis burial depths of 10, 15, 20, and 25 m, the influence of surface loading on longitudinal deformation under different depths is studied. The elastic modulus of the soil is taken as 15 MPa; the surcharge size is set at 80 kPa, equivalent to a soil pile height of approximately 4 m; and the loading area is 20 m × 40 m (aligned along and perpendicular to the tunnel axis, respectively).

Figure 2.12 shows the deformation and internal force response of the subway tunnel under surface loading at different tunnel burial depths. Figure 2.12a shows that the tunnel settlement approaches 0 at a distance of about 40 m from the load center. Therefore, the width of the tunnel settlement trough is about 80 m, four times the width of the load (along the tunnel axis). In other words, when surface loading occurs, it is essential to enhance tunnel monitoring within the settlement trough width range. Figure 2.12b shows that as the tunnel burial depth decreases from 25 to 10 m, the maximum tunnel settlement under surface loading increases from 8 to 21 mm, an approximately twofold increase in the maximum settlement. This means that under surface loading, shallow-buried tunnels exhibit significant deformation responses, which require close attention from relevant personnel.

Figure 2.12c, d show the responses of the bending moment and joint opening width of the tunnel segment rings under surface loading. Since the opening width is calculated using Eq. (2.13), the distribution pattern of the bending moment and opening width along the tunnel axis is completely consistent. The maximum positive bending moment occurs at the section of the segments corresponding to the load center, while the maximum negative bending moment occurs at the tunnel segments located 1.25 times the load width away from the load center. Figure 2.12e–f show the response of the shear force in segments and misalignment under surface loading. Since the segment misalignment is calculated using Eq. (2.16), the distribution pattern of the shear force and misalignment along the tunnel axis is completely consistent. Under surface loading, the maximum shear force in segments occurs at the section corresponding to the edge of the load. Consequently, the maximum segment misalignment also occurs in this section.

The segment rings experience a certain amount of opening width and misalignment under the effect of bending moments and shear load. When the joint opening width > 6 mm, or when the misalignment > 10 mm, the sealing gasket of circumferential joints will fail, leading to risks, such as water leakage, mud leakage, and

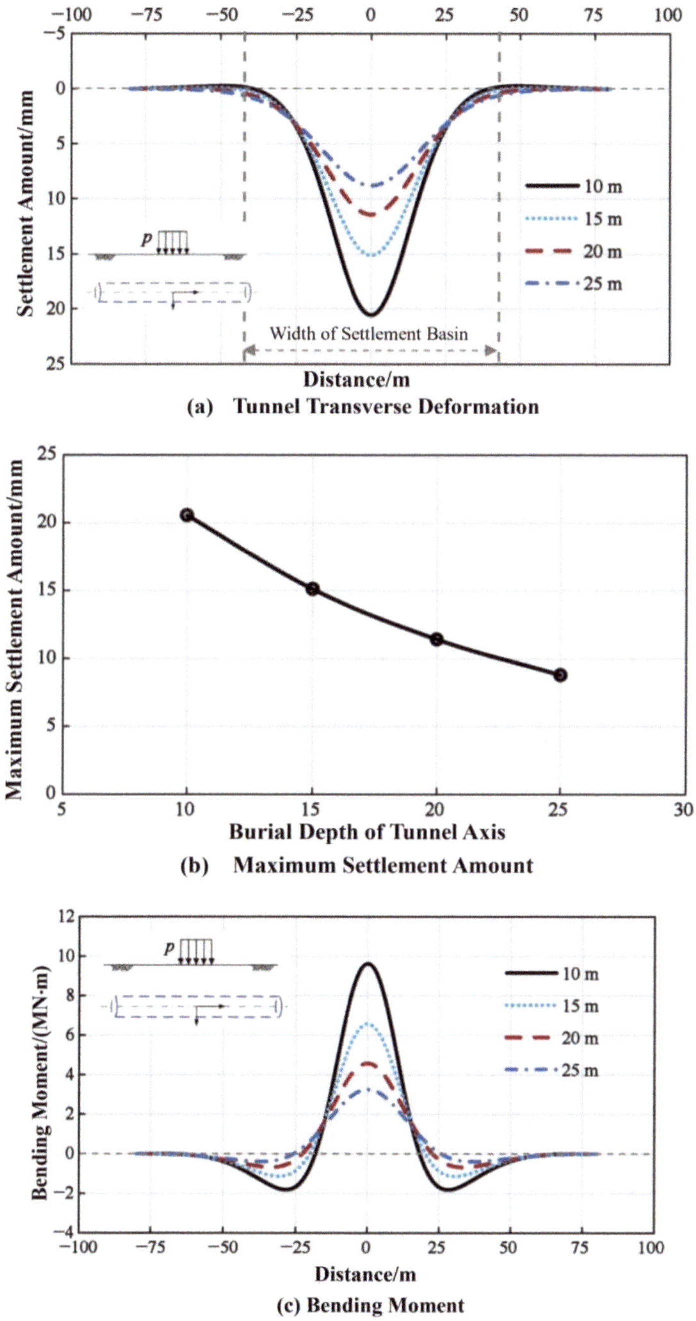

Fig. 2.12 Deformation and internal force response of tunnels at different burial depths

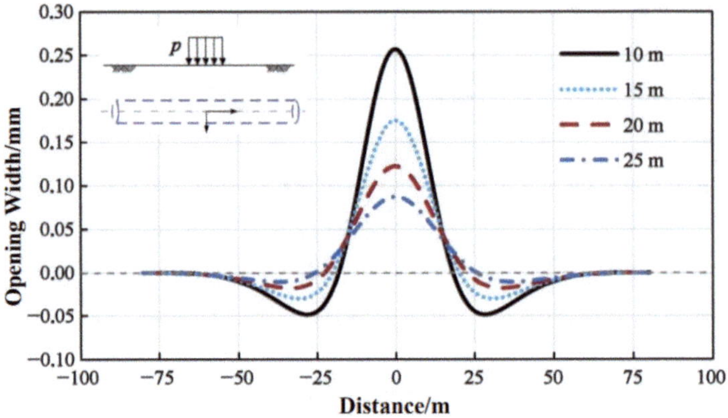

(d) Segment Opening Width

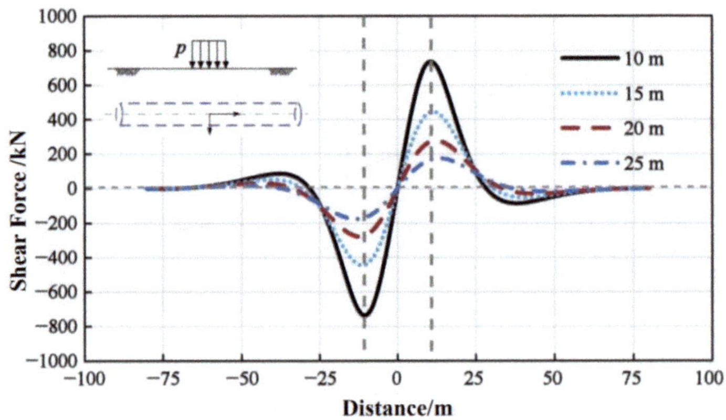

(e) Shear Force

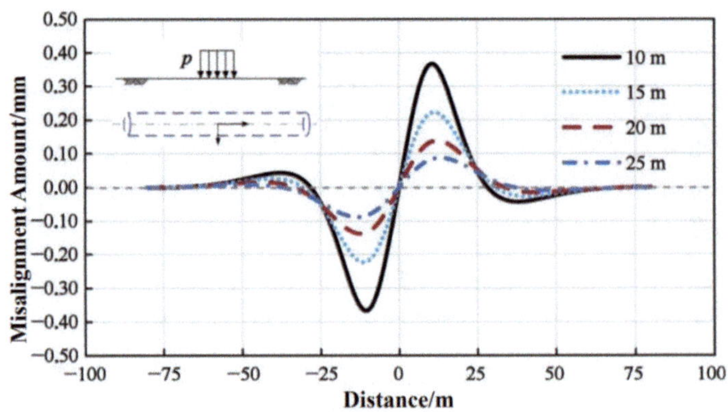

(f) Misalignment Amount

Fig. 2.12 (continued)

segment damage. Therefore, when loading occurs above the tunnel, it is necessary to enhance the detection and monitoring of these sections to prevent structural damage to the tunnel and maintain normal operation.

In summary, at the center and edge of the loading, the tunnel is subjected to large bending moments and shear forces and then experiences significant opening width and misalignment. In severe cases, this may lead to other diseases, such as trackbed detachment, crushing of segment concrete, bolt shearing, water leakage, and longitudinal joint openings, all of which threaten the safe operation of the tunnel. Therefore, when a surcharge occurs above the tunnel, operators shall immediately unload and monitor the tunnel cross-sections to prevent further severe damage.

2.3.2 Loading Magnitude

With surface loading magnitudes set to 20, 40, 60, 80, and 100 kPa, this section aims to study the effects of different loading magnitudes on longitudinal deformation. The tunnel axis burial depth is set to 10 m, with other parameters being the same as described in Sect. 2.3.1.

The longitudinal deformation under different surface loading magnitudes is shown in Fig. 2.13a, b. When the loading increases from 20 to 100 kPa, the maximum settlement increases from 5 to 26 mm, indicating that a fourfold increase in loading results in a 4-fold increase in tunnel settlement. Since the soil follows an elastic constitutive model, the maximum settlement increases linearly with the loading magnitude. Once the surface loading exceeds 80 kPa, the maximum settlement exceeds the code-specified limit of 20 mm. If the nonlinear characteristics of the circumferential joints and soil stiffness are considered, the tunnel settlement would exceed the values calculated and shown in the figures.

The internal force responses under different loading magnitudes are shown in Fig. 2.13c–f. As the loading increases, the internal forces in segments also increase. The trend of this change is similar to that described in Sect. 2.3.1.

2.3.3 Loading Range

Assume that the length of the surface loading along the tunnel axis (L_2) is fixed at 20 m and that the length perpendicular to the tunnel axis (L_1) is set to 5, 10, 20, 40, and 80 m (L_1/L_2 ratios of 0.25, 0.5, 1, 2, and 4, respectively). The impact of loading range on the longitudinal deformation is studied. Other parameters are consistent with those described in Sect. 2.3.1.

The responses of tunnel longitudinal deformation under different loading ranges are shown in Fig. 2.14a, b. As the loading range expands, the longitudinal settlement also increases. However, unlike the cases in Sects. 2.3.2 and 2.3.3, when the loading ratio $L_1/L_2 > 2$, expanding L_1 has a diminishing effect on the longitudinal

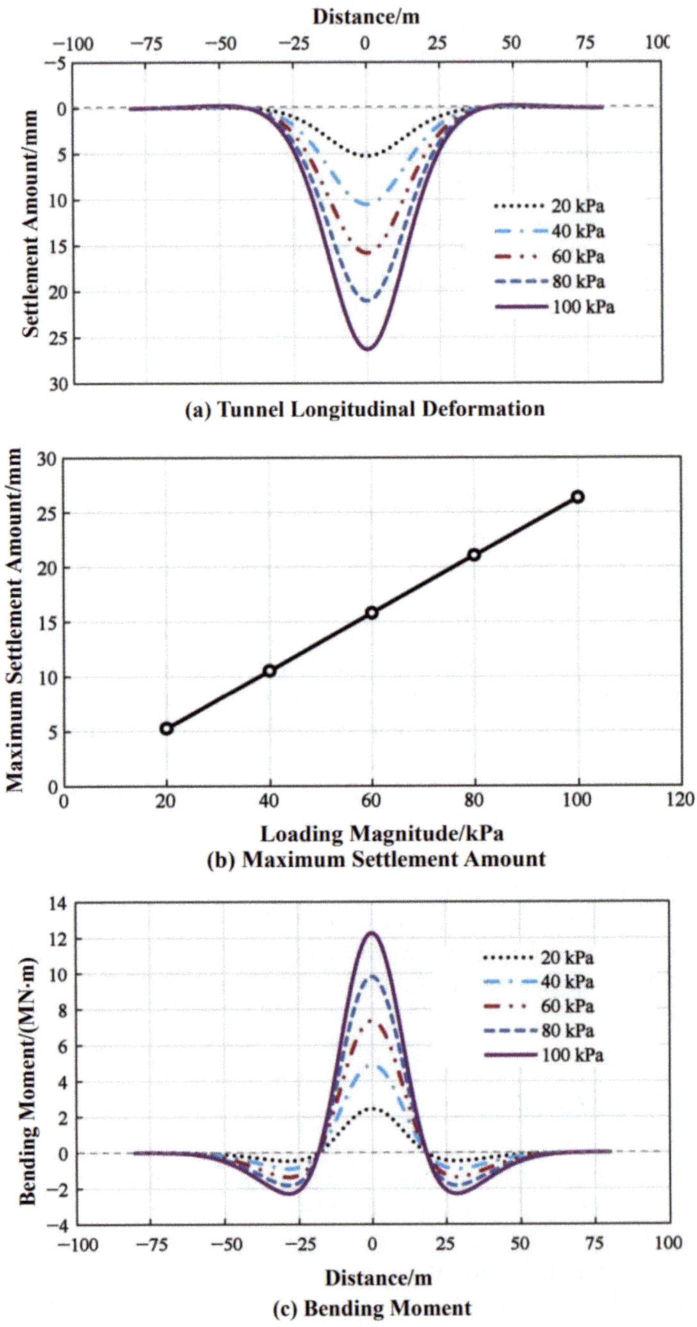

Fig. 2.13 Tunnel deformation and internal force responses under different loading magnitudes

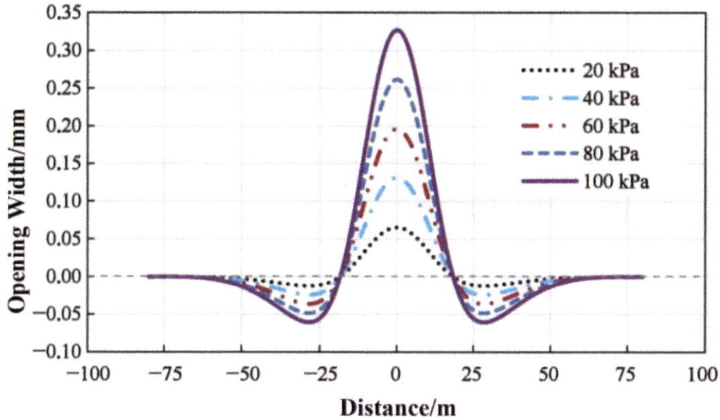

(d) Segment Opening Width

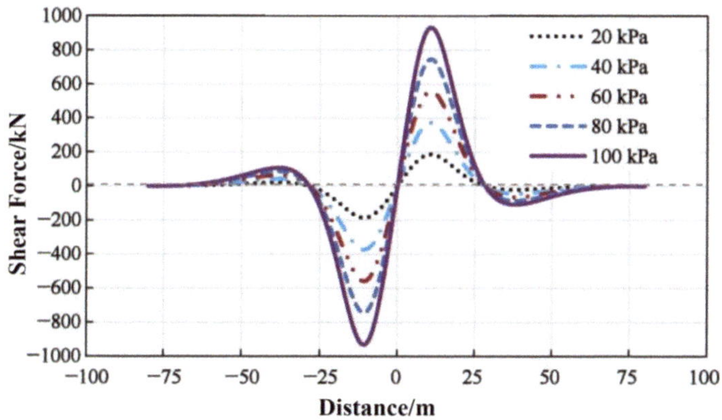

(e) Shear Force

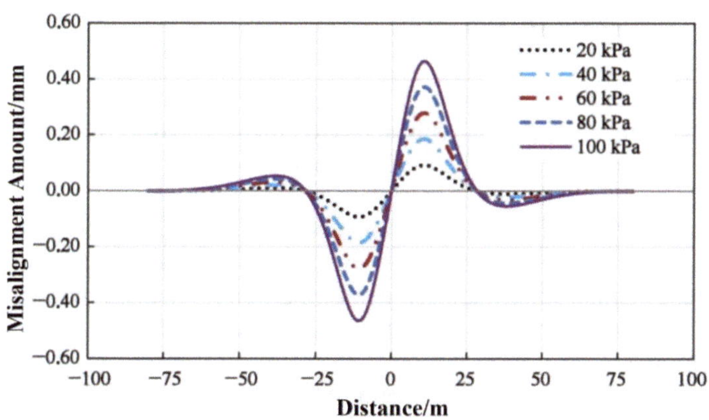

(f) Misalignment Amount

Fig. 2.13 (continued)

differential settlement. The maximum bending moment, shear force, opening width, and misalignment of segments change with the loading, similar to the maximum settlement.

2.3.4 Loading Eccentricities

In this chapter, loading eccentricity refers to the distance from the edge of the load to the tunnel axis, indicating the spatial relationship between loading and tunnel. To study the impact of surface loading under different eccentricities on longitudinal deformation, the surface loading eccentricities are set to −3.2, −1.2, −0.2, 0.7, and 1.8 D_t, and the depth of the tunnel axis is taken as 10 m, with other parameters consistent with those described in Sect. 2.3.1.

Figure 2.15 shows that when the eccentricity increases, both the tunnel deformation and the segment internal force tend to decrease. To be specific:

① When the eccentricity is less than two times the outer diameter of the tunnel, the effect of surface loading on longitudinal settlement is more significant;
② When it is greater than two times and less than four times, the effect is limited;
③ When it is greater than four times, the effect is relatively small.

Given the actual tunnel burial depths, an eccentricity of twice the outer diameter indicates that the loading is outside the main failure surface. Therefore, to minimize the impact of surface loading on the tunnel, it is recommended to keep the loading eccentricity outside the main failure surface as much as possible.

2.3.5 Elastic Modulus of Soils

To study the effect of surface loading on longitudinal deformation under different soil elastic moduli, the elastic moduli are set to 5, 10, 15, and 20 MPa, and the tunnel axis depth is taken as 10 m, with other parameters the same as those described in Sect. 2.3.1.

The tunnel longitudinal deformation and the internal force response under different soil elastic moduli are shown in Fig. 2.16. As the elastic modulus increases, the tunnel longitudinal deformation gradually decreases. When the modulus increases from 5 to 20 MPa, the maximum tunnel settlement decreases from 56 to 15 mm, representing a 70% reduction. However, the rate of decrease in settlement diminishes. Therefore, under surface loading, it is essential to thoroughly investigate the soil properties within the range of tunnel burial depth. If the soil elastic modulus is below 15 MPa, the tunnel longitudinal deformation under loading may exceed the code-specified limits. Additionally, as the modulus decreases, the deformation within the tunnel settlement basin increases significantly, which differs markedly from the results obtained for other parameters.

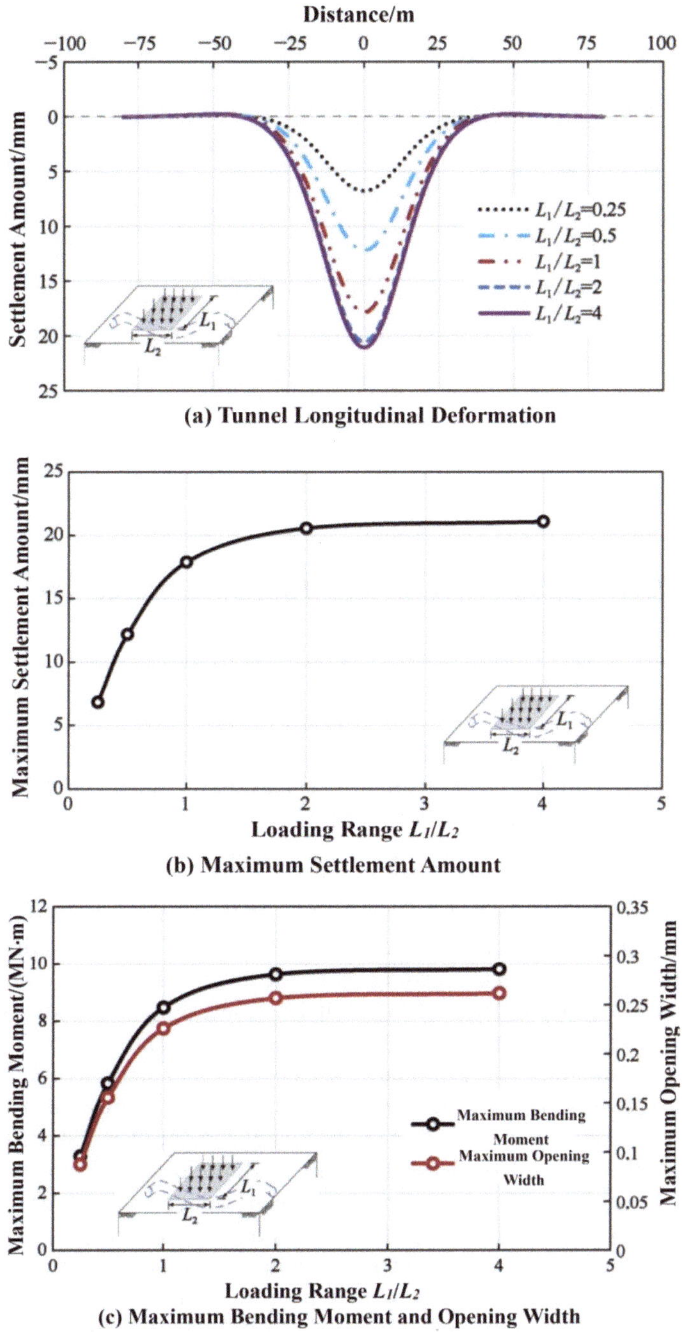

Fig. 2.14 Tunnel deformation and internal force response under different loading ranges

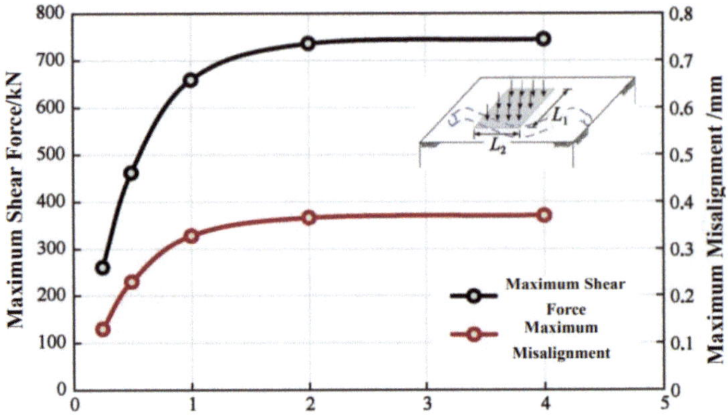

(d) Maximum Shear Force and Misalignment

Fig. 2.14 (continued)

2.3.6 Effects of Soil Properties on Tunnel Longitudinal Deformation

(1) Overlying Soil Layer

Assume the tunnel axis depth z_0 is 15 m, the surface loading *p* is 72 kPa, and the loading range is 20 m × 40 m. For simplified analysis, assume the first layer is 10 m thick, the elastic modulus varies from 5 to 40 MPa, and the Poisson's ratio is 0.35. The second layer is a semi-infinite space body with an elastic modulus of 10 MPa and the Poisson's ratio of 0.35.

Under the loading above the tunnel, the effect of the modulus of the overlying soil layer on the longitudinal deformation is shown in Fig. 2.17a. The maximum longitudinal settlement occurs at the loading center. As the modulus of the overlying soil layer increases, the longitudinal deformation decreases. When the modulus increases from 5 to 40 MPa, the maximum tunnel settlement decreases from 16 to 12 mm, representing a 25% reduction. This is primarily because a higher modulus of the overlying soil layer indicates greater soil strength and more pronounced stress diffusion effects. Therefore, the additional stress acting on the tunnel is reduced, resulting in a smaller settlement of the tunnel. Near 40 m from the loading center, the tunnel settlement is almost 0. Therefore, the width of the settlement trough is about 80 m, which is four times the loading width (along the tunnel axis). Additionally, the width of the settlement trough is nearly unaffected by the modulus of the overlying soil layer.

Figure 2.17b, c show the changes in bending moment and opening width with different moduli of the overlying soil layer. The figures show that under loading, the bending moment and opening width have a consistent distribution along the tunnel's longitudinal direction. The maximum values occur at the loading center, which is the most critical section of the tunnel. When tunnel segments are exposed to

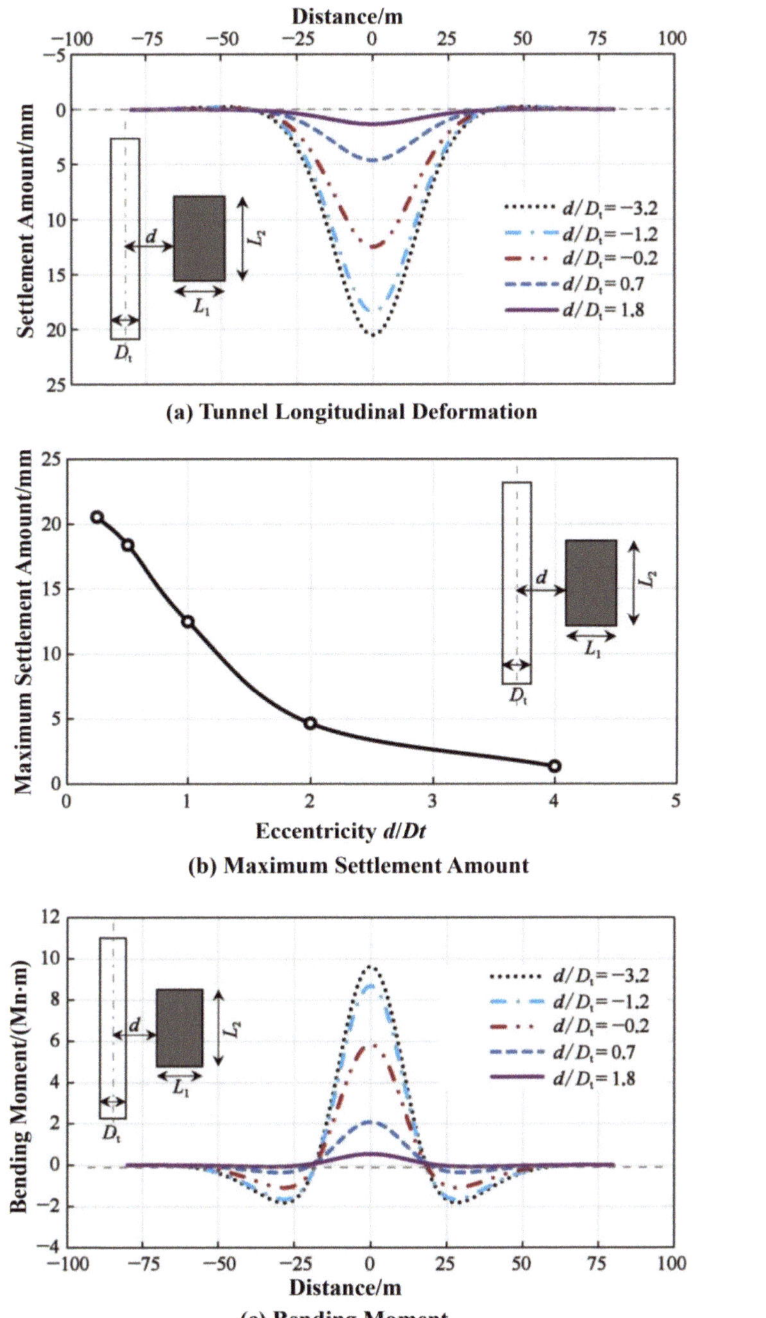

Fig. 2.15 Tunnel deformation and internal force response under different loading eccentricities

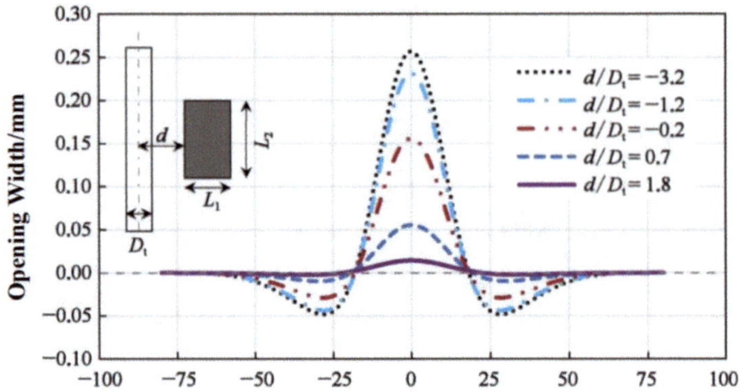

(d) Segment Opening Width

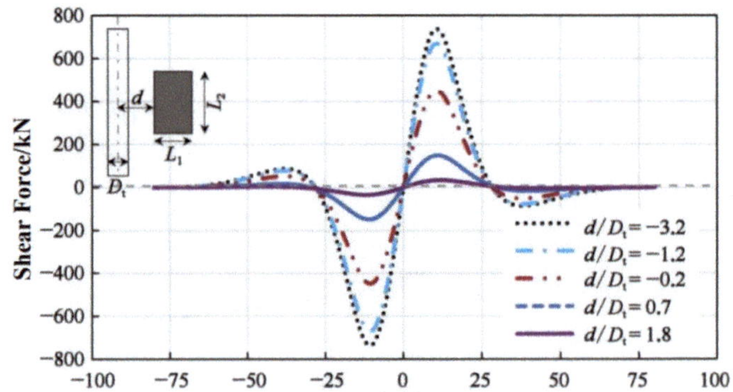

(e) Shear Force

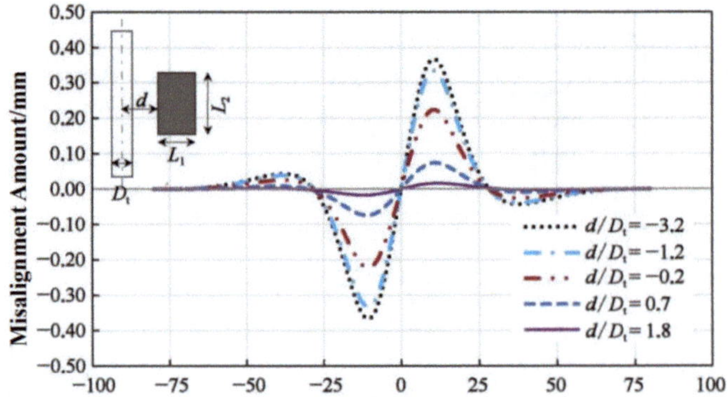

(f) Misalignment Amount

Fig. 2.15 (continued)

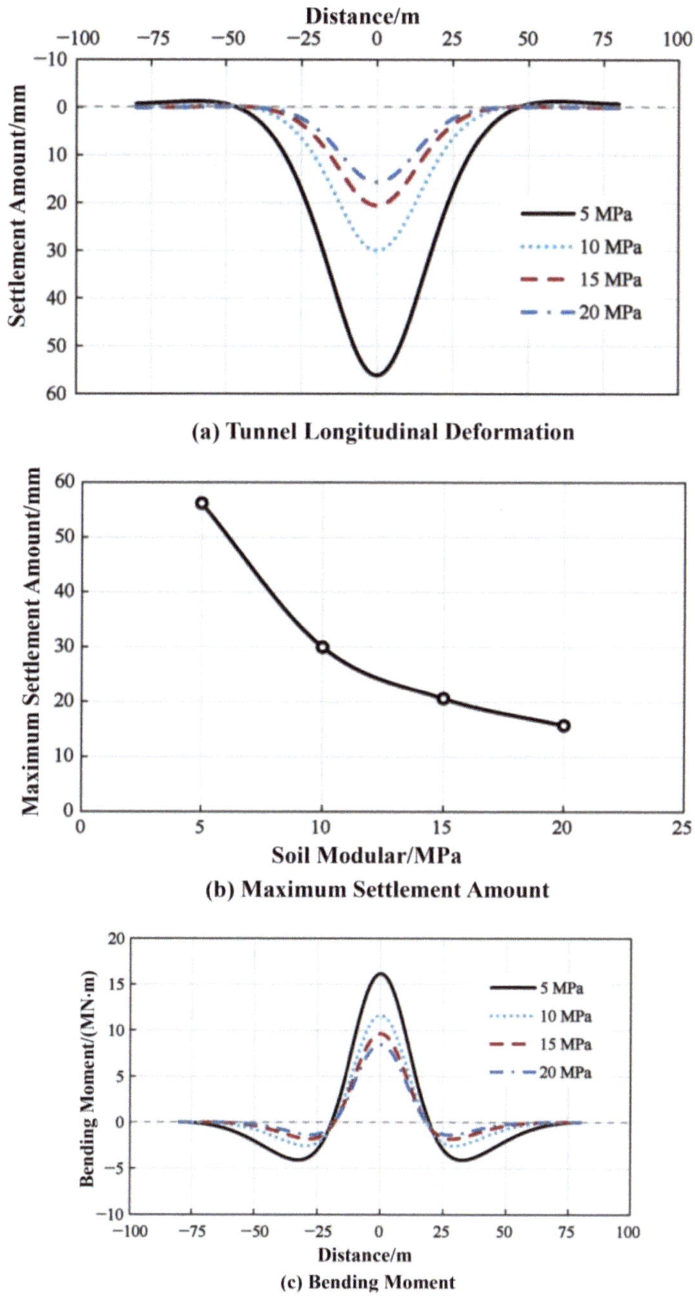

Fig. 2.16 Tunnel deformation and internal force response under different soil moduli

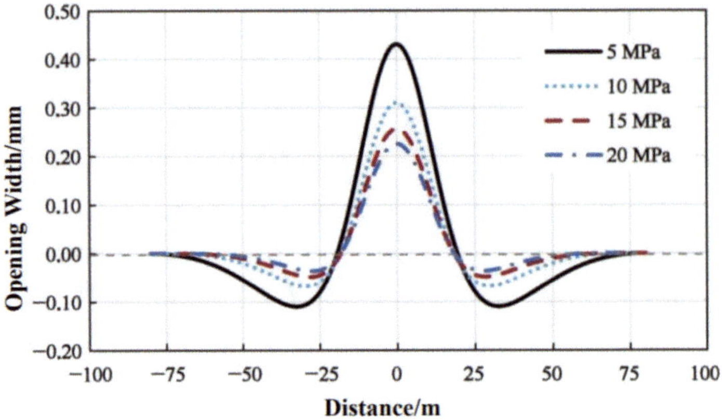

(d) Segment Opening Width

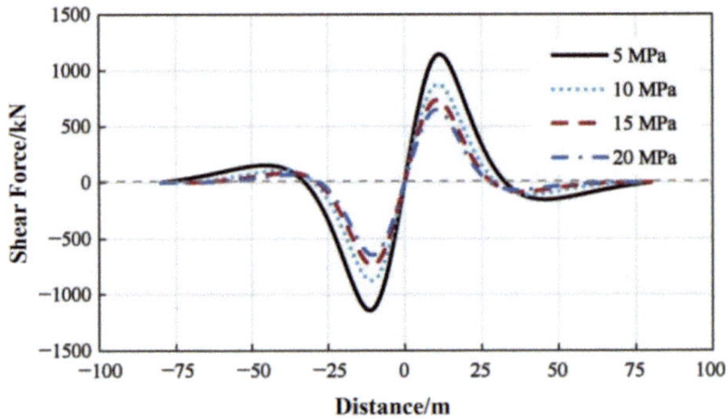

(e) Shear Force

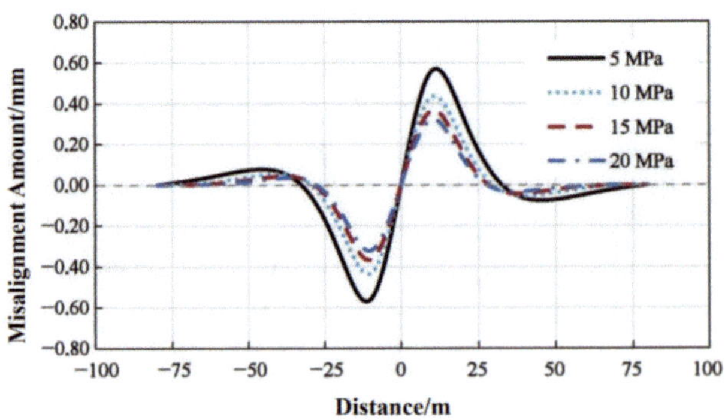

(f) Misalignment Amount

Fig. 2.16 (continued)

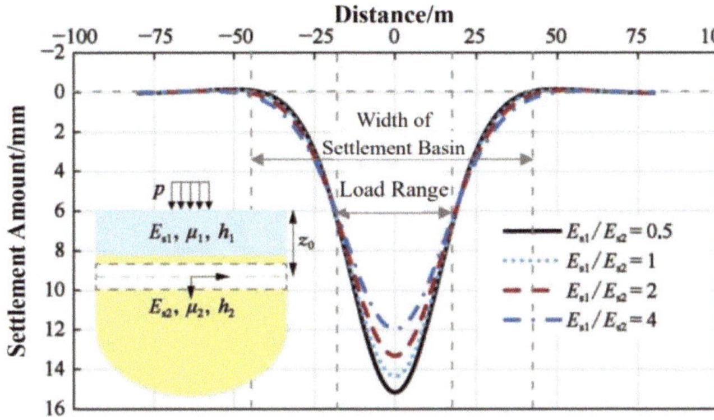

(a) Influence of Overlying Soil Modulus on Tunnel Settlement

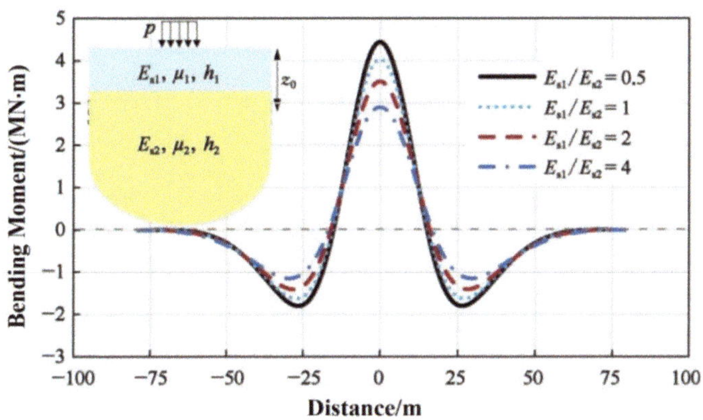

(b) Influence of Overlying Soil Modulus on Tunnel Bending Moment

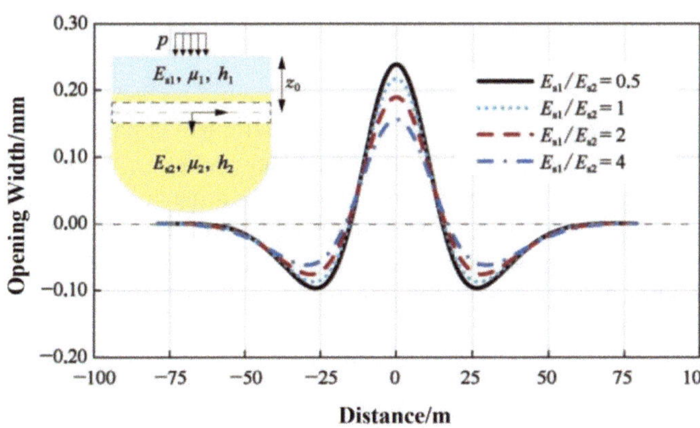

(c) Influence of Overlying Soil Modulus on Tunnel Opening Width

Fig. 2.17 Influence of the modulus of the overlying soil layer on tunnel internal forces and deformation

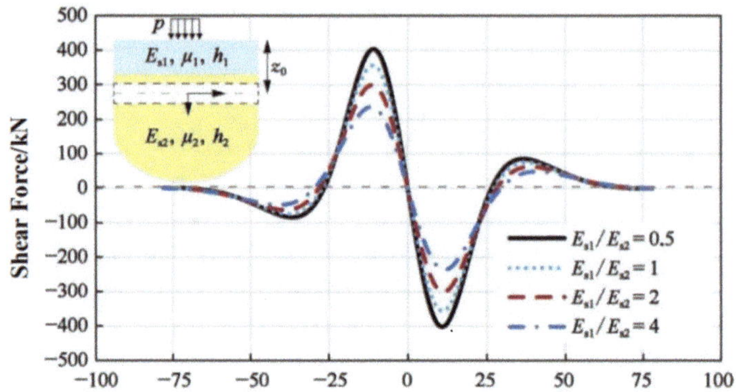

(d) Influence of Overlying Soil Modulus on Tunnel Shear Force

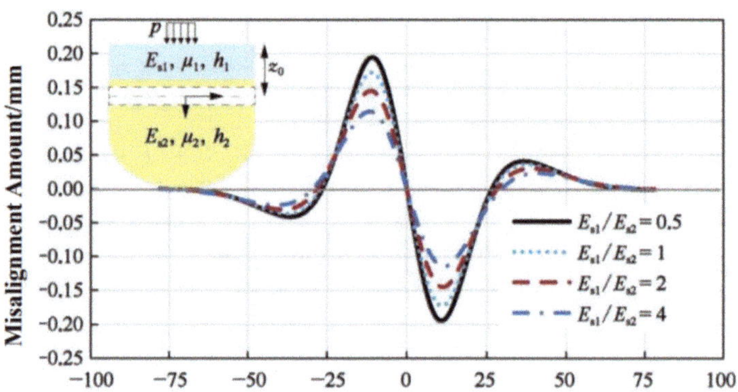

(e) Influence of Overlying Soil Modulus on Tunnel Misalignment

Fig. 2.17 (continued)

excessive compressive or tensile stresses, they experience damage, cracks, or water leakage, which leads to water infiltration and endangers the tunnel operation. Under surface loading, the maximum shear force and misalignment occur at the edge of the loading, as shown in Fig. 2.17d, e. This indicates that at these tunnel sections, the bolts experience the maximum shear stress, and an increase in misalignment between tunnel rings makes them more prone to shear failure.

(2) Underlying Soil Layer

For simplified analysis, assume the first layer is 20 m thick, with an elastic modulus of 10 MPa, and the Poisson's ratio of 0.35; the second layer is a semi-infinite space body with elastic modulus varying from 5 to 40 MPa and the Poisson's ratio of 0.35.

Under the surface loading above the tunnel, the impact of the underlying soil modulus on tunnel internal forces and deformations is shown in Fig. 2.18. Figure 2.18 (a) shows that the higher the modulus of the underlying soil layer is, the smaller the tunnel settlement becomes. When the modulus increases from 5 to 40 MPa, the maximum settlement decreases from 22 to 7 mm, representing a 68% reduction. This is mainly because a higher modulus represents higher soil stiffness and imposes more constraints on the overlying soil layer. When the modulus approaches infinity, the underlying layer can be considered as a rigid boundary, effectively constraining the vertical displacement of the bottom of the overlying soil. In this case, the longitudinal deformation entirely depends on the properties of the upper soil layer.

Moreover, compared to Fig. 2.18a, as the modulus increases, the settlement range significantly decreases, from 5 times the loading width to 3 times.

The influence of the modulus of the underlying soil layer on the bending moment, opening width, shear force, and misalignment is shown in Fig. 2.18b–e, respectively. As the modulus increases, the internal forces and deformation decrease, following a similar trend to that of the overlying soil layer.

For two-layer soil systems, the calculation result of the ratio of the maximum tunnel settlement to the soil modulus is shown in Fig. 2.19. The results demonstrate that under surface loading, increasing the modulus of either the overlying or underlying soil layers reduces the maximum longitudinal settlement. By comparison, the effect on the underlying soil layer is more pronounced. It shows that to control tunnel settlement, better results can be achieved by controlling the underlying soil layer rather than the hard surface foundation. Furthermore, when the modulus of the underlying soil layer is below that of the soil layer where the tunnel is located (i.e., an underlying soft soil layer), the tunnel settlement increases dramatically. Therefore, special attention should be paid when the tunnel is underlain by soft soil and surface loading is applied above the tunnel. The piling shall be avoided to prevent an excessive deformation caused by the surcharge.

2.3.7 *Safety Factor Evaluation of Tunnel Longitudinal Deformation Under Surcharge Loading*

The safety factor for tunnel longitudinal settlement under surcharge loading can be defined as

$$F_{sv} = S_{lit}/S,$$

where, S_{lit} is the specified limit value of 20 mm and S is the maximum settlement under surcharge. Under different burial depths of the tunnel axis, F_{sv} varies with the loading magnitude and soil elastic modulus, as shown in Fig. 2.20 (with the loading range set to 20 m × 40 m).

F_{sv} increases as the burial depth and soil modulus increase and decreases as the loading magnitude increases. For example, with a soil modulus of 10 MPa, when the burial depth is 10 m, the surface surcharge should not exceed 50 kPa; when the tunnel burial depth is 20 m, the maximum surcharge can reach 90 kPa. It should be noted

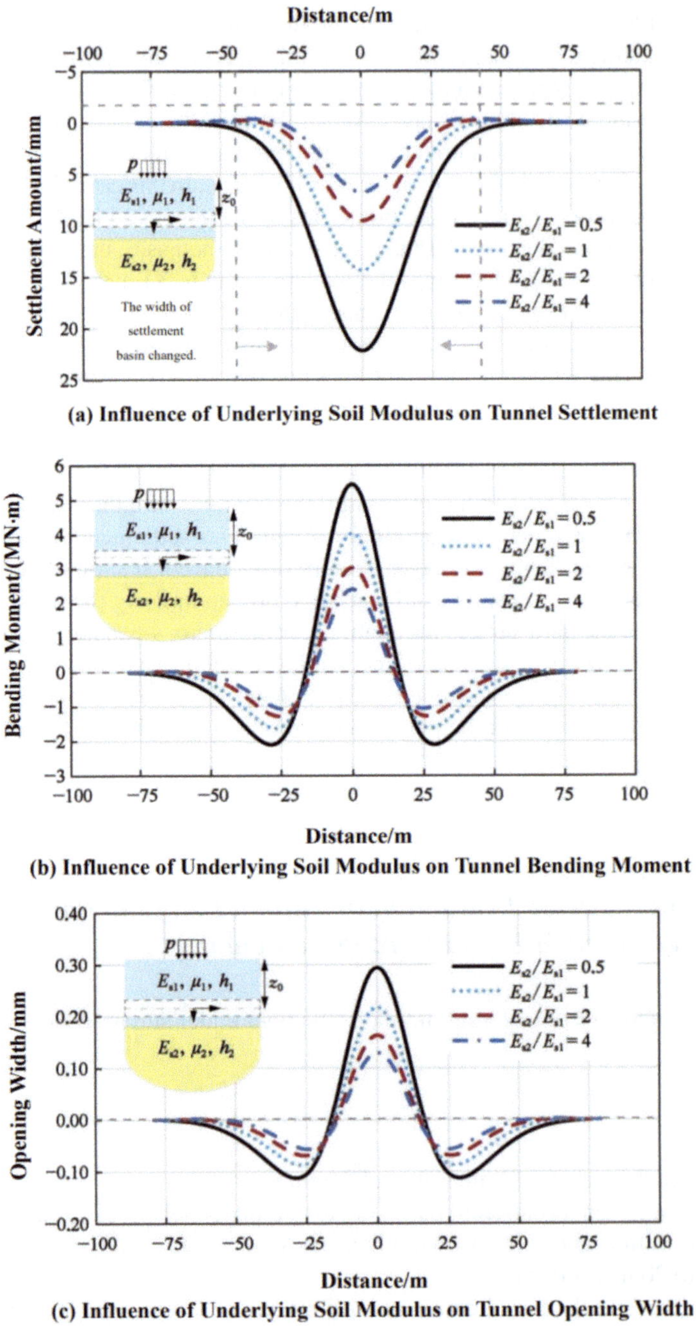

Fig. 2.18 Influence of the modulus of the underlying soil layer on tunnel internal forces and deformations

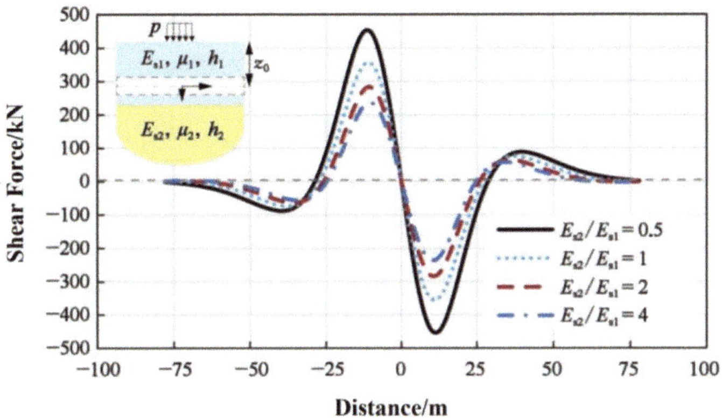

(d) Influence of Underlying Soil Modulus on Tunnel Shear Force

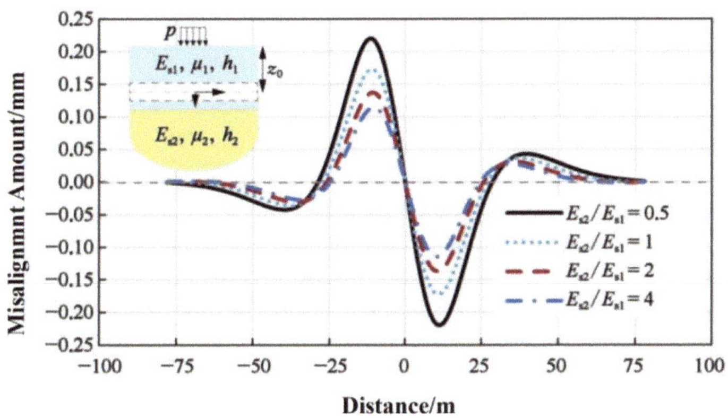

(e) Influence of Underlying Soil Modulus on Tunnel Misalignment

Fig. 2.18 (continued)

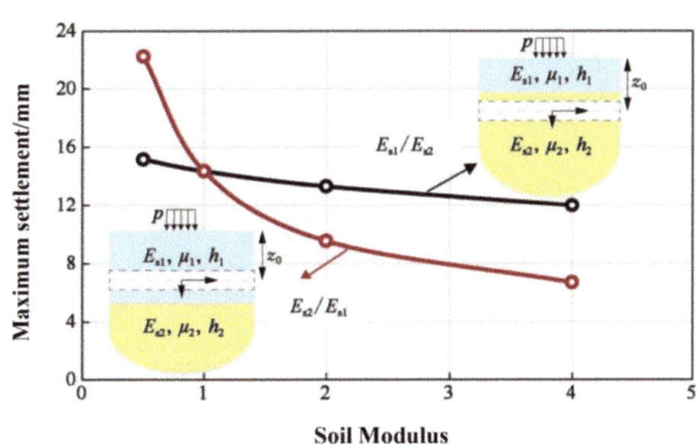

Fig. 2.19 Variation of maximum settlement with soil modulus ratio

Fig. 2.20 Contour map of safety factor F_{SV}

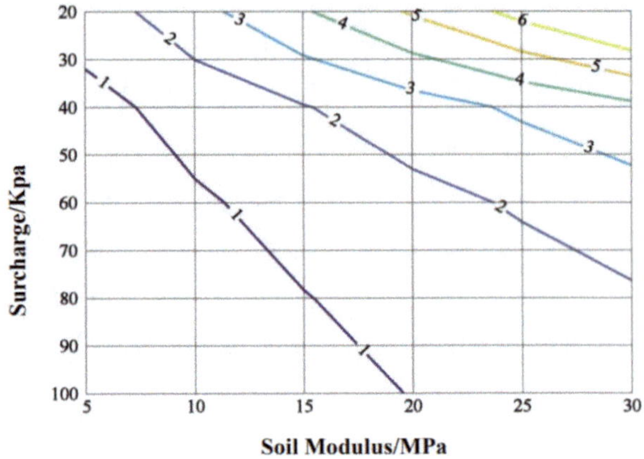

(a) Tunnel axis at 10m

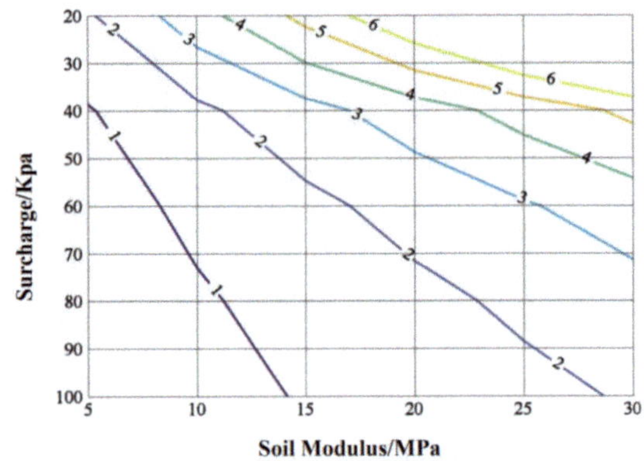

(b) Tunnel axis at 15m

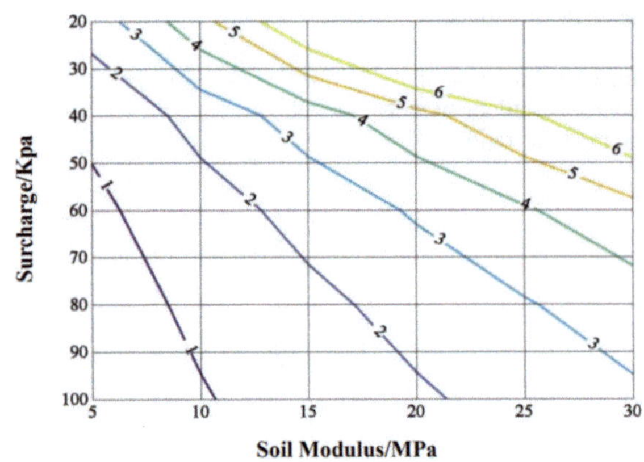

(c) Tunnel axis at 20m

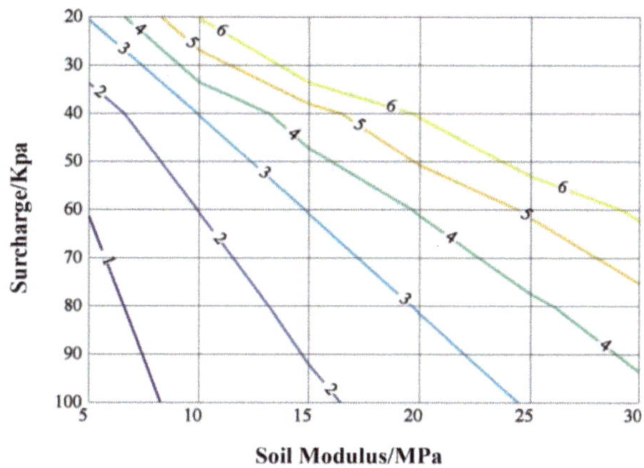

Fig. 2.20 (continued)

(d) Tunnel axis at 25m

that the calculations above do not consider the initial longitudinal deformation, so further verification is required in practical engineering projects.

2.4 Summary

In this chapter, the tunnel is modeled as a Timoshenko beam, and the Pasternak foundation model is used to establish a theoretical calculation model for the response of tunnel longitudinal deformation under surface loading. The focus is on analyzing the effects of tunnel burial depth, loading magnitude, loading range, loading eccentricity, soil modulus, and the layered characteristics of soils on tunnel longitudinal deformation. The main conclusions are:

① Due to the longitudinal structural characteristics, tunnel's longitudinal deformation mainly consists of bending deformation and misalignment, which can be effectively captured by the Timoshenko beam model.

② Tunnel burial depth, loading magnitude, loading range, and soil modulus significantly affect the tunnel longitudinal deformation. Specifically, for subway tunnels in Shanghai, when the elastic modulus of the soil layer surrounding the tunnel is below 15 MPa, the longitudinal deformation under surface loading may exceed the code-specified limits. Thus, a strict evaluation is required.

③ As the loading eccentricity increases, the tunnel longitudinal deformation under surface loading decreases. When the loading eccentricity is outside the primary failure surface, the influence of loading on the longitudinal settlement deformation can be negligible (though reassessment is necessary when the loading is significant).

④ The elastic modulus of both overlying and underlying soil layers significantly influences the tunnel settlement and internal forces. Increasing the elastic modulus of the underlying soil layer can significantly reduce tunnel settlement. When the tunnel is underlain by soft soil layers, it is more prone to longitudinal deformation under surface loading. Therefore, to ensure the safe operation of the tunnel, surface loading above the tunnel shall be prohibited.

References

Ai ZY, Wu C (2008) Transfer matrix solution for layered foundation in three-dimensional Cartesian coordinate system. J Chongqing Jianzhu Univ 30(02):43–46

Dai HW, Chen RP, Chen YM (2006) Analysis of new surface construction loads on longitudinal deformation of adjacent metro tunnels. Chin J Geotechn Eng 28(03):312–316

Fan YY, Guo XH, Deng ZJ et al (2014) Impact analysis of overlying loading/unloading on shield tunnel deformation. Constr Technol 43(07):107–109

Huang M, Zhang C, Li Z (2009) A simplified analysis method for the influence of tunneling on grouped piles. Tunn Undergr Space Technol 24(04):410–422

Kerr A (1985) On the determination of foundation model parameters. J Geotechn Eng 111(11):1334–1340

Klar A, Vorster T, Soga K et al (2005) Soil-pipe interaction due to tunnelling: comparison between Winkler and elastic continuum solutions. Geotechnique 55(06):461–466

Li JY, Qian JG, Ru ZM et al (2015) Three-dimensional analysis method for influence of high fill on buried pipelines in layered foundation. Chin J Geotech Eng 37(S2):65–69

Li P, Du S, Shen S et al (2016) Timoshenko beam solution for the response of existing tunnels because of tunneling underneath. Int J Numer Anal Methods Geomech 40:766–784

Liang FY, Li YC, Huang MS (2013) Simplified analysis method for horizontal piles based on Pasternak two-parameter foundation model. Chin J Geotechn Eng 35(S1):300–304

Liang R, Xia T, Huang M et al (2017) Simplified analytical method for evaluating the effects of adjacent excavation on shield tunnel considering the shearing effect. Comput Geotech 81(01):167–187

Liang R, Wu W, Yu F (2018) Simplified method for evaluating shield tunnel deformation due to adjacent excavation. Tunn Undergr Space Technol 71(01):94–105

Liang FY, Yuan Q, Li JP, Zhang SX (2020) Influence of soil stratification characteristics on longitudinal deformation of metro tunnels under surcharge loading. Chin J Geotechn Eng 42(01):63–71

Liang F, Yuan Q, Song Z et al (2021) Longitudinal responses of shield tunnel subjected to surcharge considering dislocation. Proc Inst Civ Eng Geotech Eng 174(4):342–354

Peng W, Zhao M, Zhao H et al (2020) A two-pile foundation model in sloping ground by finite beam element method. Comput Geotech 122(06):103503

Shen S, Wu H, Cui Y et al (2014) Long-term settlement behaviour of metro tunnels in the soft deposits of Shanghai. Tunn Undergr Space Technol 40(02):309–323

Shiba Y, Kawashima K, Obinata N et al (1988) An evaluation method of longitudinal stiffness of shield tunnel linings for application to seismic response analyses. Doboku Gakkai Ronbunshu **398**, 319–327 [in Japanese]

Tanahashi H (2004) Formulas for an infinitely long Bernoulli-Euler beam on the Pasternak model. Soils Found 44(05):109–118

Vesic AB (1961) Bending of beams resting on isotropic elastic solid. J Eng Mech Div 87(02):35–54

Wang RL (2009) Analysis of longitudinal deformation of Shanghai metro shield tunnels. Undergr Eng Tunn 4:1–7

Wu H, Shen S, Liao S et al (2015) Longitudinal structural modelling of shield tunnels considering shearing dislocation between segmental rings. Tunn Undergr Space Technol 50(08):317–323
Xu L (2005) Study on longitudinal settlement of shield tunnels in soft soil. PhD dissertation, Tongji University, Shanghai
Ye F, He C, Zhu HH et al (2011) Analysis of longitudinal equivalent stiffness of shield tunnel considering transverse performance. Chin J Geotechn Eng 33(12):1870–1876
Yu J, Zhang CR, Huang MS (2012) Foundation modulus of buried pipelines in passive state. Chin J Rock Mech Eng 31(01):123–132
Yu J, Zhang C, Huang M (2013) Soil-pipe interaction due to tunnelling: assessment of Winkler modulus for underground pipelines. Comput Geotechn 50(05):17–28
Zhang D, Huang Z, Li Z et al (2019a) Analytical solution for the response of an existing tunnel to a new tunnel excavation underneath. Comput Geotech 108(04):197–211
Zhang C, Zhu J, Huang M et al (2019b) Winkler load-transfer analysis for pipelines subjected to surface load. Comput Geotech 111(07):147–156

Chapter 3
Tunnel Transverse Deformation Under Surcharge Loading: Analysis and Model Tests

3.1 Introduction

The deformation and structural performance of subway shield tunnels under adverse top overloading conditions can be generally analyzed in both longitudinal and transverse directions. In terms of longitudinal deformation, the primary manifestation is the differential settlement caused by joint dislocation; whereas transverse deformation is characterized by the transformation from a circular to a transversely elliptical cross-section, often accompanied by longitudinal joint opening or closing.

However, in actual shield tunnel engineering, excessive transverse convergence deformation frequently leads to incidents such as water seepage, bolt failure, and tube sheet cracking, which pose a significant threat to the safety of subway tunnels. Therefore, it is necessary to analyze the mechanism of transverse deformation damage in subway shield tunnels under loading.

This chapter explores the transverse deformation mechanism of subway shield tunnels under ground loading and the specific impacts of various related parameters through a combination of finite element numerical analysis and indoor scaled-down model tests.

R. Wang et al., *Impact Analysis and Control Strategies of Overlying Loading/Unloading Effects on Soft Soil Shield Tunnels*, Key Technologies for Tunnel Construction under Complex Geological and Environmental Conditions,
https://doi.org/10.1007/978-981-95-1462-5_3

3.2 Analytical Model for Transverse Deformation of Tunnels

3.2.1 Finite Element Model

Based on different assumptions regarding the interaction between the ground and the tunnel structure, analysis for lateral deformation of shield tunnels can be categorized into two primary approaches: the "load-structure method" and the "stratigraphic structure method" (Liu and Hou 1991). The load-structure method (Fig. 3.1a) assumes that the principal role of the ground following tunnel excavation is to exert loads on the tunnel structure, which must be capable of safely and reliably withstanding these loads, including ground pressure. In this method, the ground pressure is initially determined using stratigraphic classification methods or practical formulas, after which the internal forces and deformations of the tunnel are computed employing the calculation method for structures on elastic foundations. By assuming the distribution forms of ground pressure and resistance, this method simplifies the interaction between the tunnel and the ground. Depending on different assumptions, it can be further divided into the (modified) conventional method, the multi-hinged circular ring model, and the beam-spring model.

The stratigraphic structure method [Fig. 3.1b] does not simplify the interaction between the tunnel structure and the surrounding ground. Instead, it treats them as a unified system. Based on deformation coordination conditions, this method calculates the internal forces and deformations of both the tunnel structure and the surrounding strata. Under loading, the interaction between the strata and the structure becomes highly complex. The load structure method, which relies on simplified assumptions,

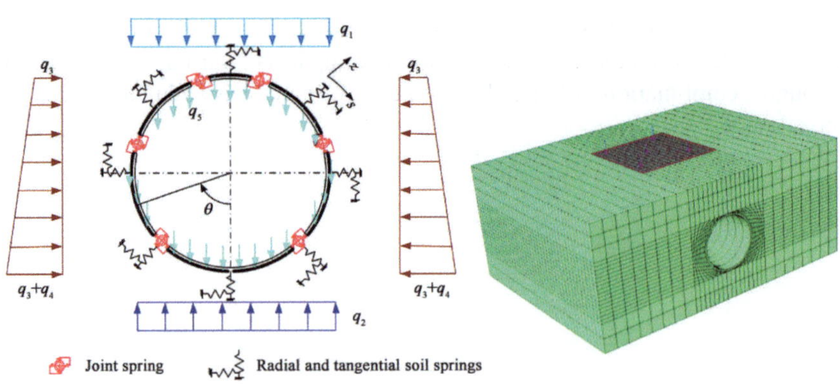

(a) The load-structure method (b) The stratigraphic structure method

Fig. 3.1 Calculation model of transverse deformation of the tunnel

may fail to accurately capture this mechanism. Consequently, the stratigraphic structure method is more appropriate for evaluating the transverse deformation and stress state of the tunnel.

The stratigraphic structure method primarily depends on finite element numerical calculations. Among numerous finite element software programs, ABAQUS stands out as a classic general-purpose finite element analysis tool. It offers a wide range of geotechnical constitutive models, hundreds of element types, and advanced coupled analysis capabilities, enabling the simulation of mechanical behaviors and working conditions of complex geotechnical media. In this study, ABAQUS is employed to investigate the mechanism by which ground loading influences the transverse deformation of shield tunnels in soft soil through 2D finite element analysis. The finite element modeling process is described in detail below.

(1) Tunnel model

The overlying large-area loading is simplified to a plane strain problem, and a 2D finite element model is employed for modeling.

Figure 3.2 (Huang et al. 2017) shows the cross section view of the shield tunnel structure of Shanghai subway with straight-joint assembly. The tunnel segments are composed of one top block (F), two adjacent blocks (L1 and L2), two standard blocks (B1 and B2), and one bottom block (D). The tube segments have an outer diameter of 6.2 m, a thickness of 0.35 m, and a ring width of 1.2 m.

The tunnel segments are precast from C55 concrete. In the numerical simulation, a bifurcated constitutive model is used (Wang and Zhang 2013) [Fig. 3.3a], which has a first-stage modulus of elasticity of 35.5 GPa and a second-stage modulus of elasticity of 592 MPa.

Figure 3.4a presents the detailed structure of the tunnel longitudinal joints, which are primarily composed of the concrete, elastic gaskets and bolts. Specifically, the intrados and extrados of these joints are provided with 8 and 4 mm gap, respectively. Each longitudinal joint is connected via two bolts, which have a diameter of 30 mm, a bolt hole diameter of 42 mm, and a grade of 8.8.

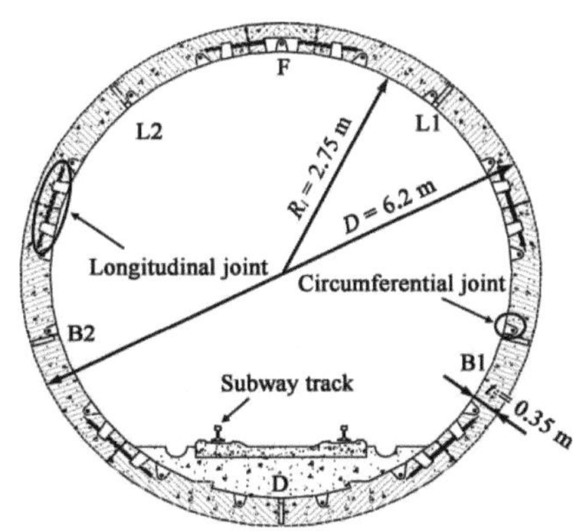

Fig. 3.2 Structure of Shanghai subway shield tunnels with straight-joint assembly

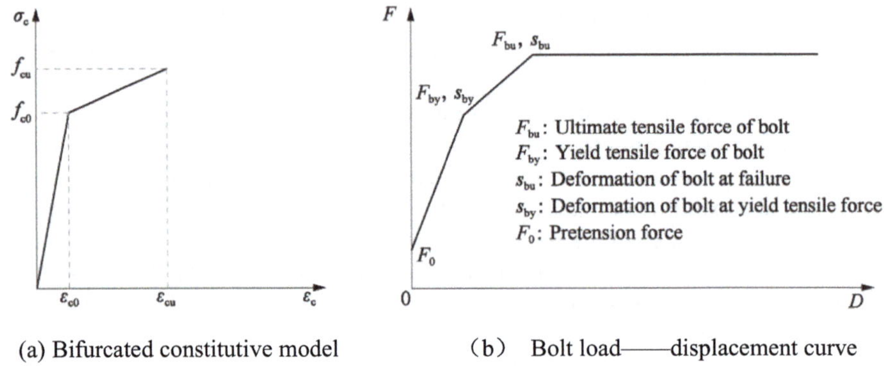

Fig. 3.3 Concrete constitutive model and bolt load—displacement curve

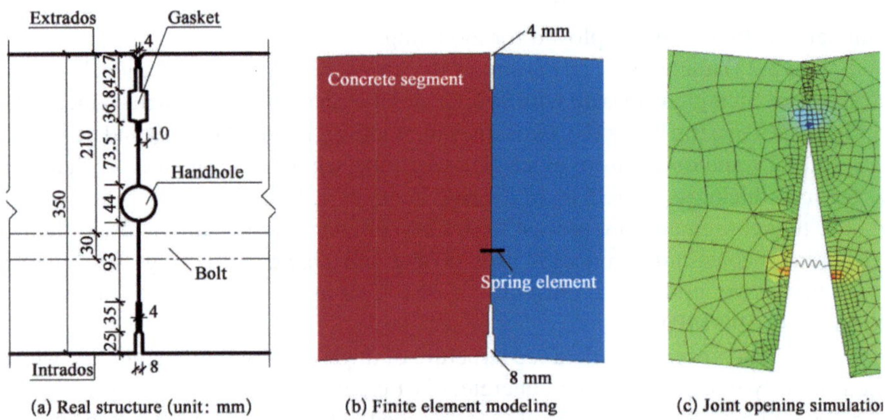

Fig. 3.4 Finite element modeling of longitudinal joints

The longitudinal joints are crucial for the transverse deformation of shield tunnels. When creating finite element models, gaps are provided at the intrados and extrados of the joints based on the actual structure, while the influence of hand holes is disregarded. Attempting to precisely model connecting bolts in numerical simulations involves complex contact coupling issues, leading to intricate models that are hard to converge and may not yield accurate results. Hence, tensile and shear springs are utilized to simulate the connecting bolts, with their nonlinear stiffness assigned to reflect the bolts' yield characteristics. Figure 3.4b illustrates the details of the lining segment's longitudinal joints in the numerical model.

The normal contact between lining segments is defined as hard contact, meaning concrete segments only transmit normal pressure when compressed, not when detached. The friction coefficient between lining segments is set at 0.4. The shear capacity of the tunnel joint is jointly provided by joint friction and shear springs.

(2) Stratigraphic model

The Shanghai area, a typical coastal region with soft soil geology, has widespread thick saturated highly compressible soft clay strata. In finite element modeling, the soil body was divided into five strata based on the actual strata distribution in Shanghai. Table 3.1 shows each layer's represented stratum and soil property parameters. The Mohr–Coulomb constitutive model was utilized to simulate the soil body.

The contact between the tunnel and soil can be simulated by surface contact, multi-point constraints (MPC), or tie constraints. This model uses surface contact. The normal direction is set as "hard contact", meaning the tunnel and soil transfer normal pressure only when compressed, not when detached. The tangential direction uses a "penalty function" to model their relative slip.

The main stratum of the subway tunnel section is mostly in the shape of a "V" slope, traversing soil strata ②–⑦. To facilitate computational analysis, Shanghai subway tunnel depths are classified into three primary categories (Fig. 3.5): ④–④, ④–⑤, and ⑤–⑤. The ④–④ type is entirely within stratum ④. The ④–⑤ type has its upper part in stratum ④ and lower part in stratum ⑤. The ⑤–⑤ type is mainly within stratum⑤ (particularly sub-stratum $⑤_1$).

The finite element analysis establishes tunnel conditions corresponding to the above three stratum conditions to analyze the influence of tunnel burial depth and stratum conditions on them.

(3) Unit selection

The tunnel and stratum are divided into quadrilateral grids. A four-node plane strain unit (CPE4) is used to simulate the soil body, while the tunnel employs a four-node plane strain incompatible unit (CPE4I) to precisely simulate bending deformation and prevent shear locking.

The finite element model, depicted in Fig. 3.6, has a width of 200 m and a depth of 60 m. A surface load with width *B* and magnitude *p* is applied to the ground surface.

Table 3.1 Parameters of soil body

	Depth range/m	Capacity/(kN/m^3)	Compression modulus/MPa	Cohesion/kPa	Angle of internal friction/(°)
①–③	0–9	18.0	3.0	12	17
④	9–18	17.0	2.3	11	12
$⑤_1$	18–28	17.6	4.8	17	19
$⑤_2$	28–38	17.6	8.0	13	25
⑥–⑦ (vi)–(vii)	38–60	19.6	10.0	5	35

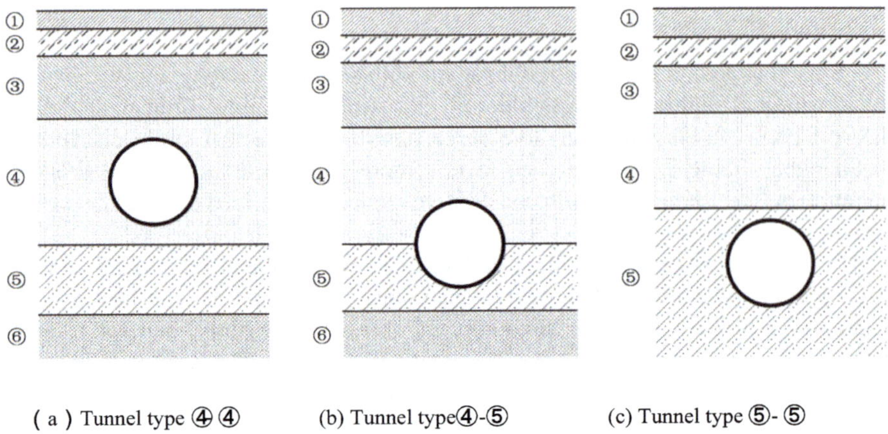

Fig. 3.5 Typical tunnel surrounding stratum

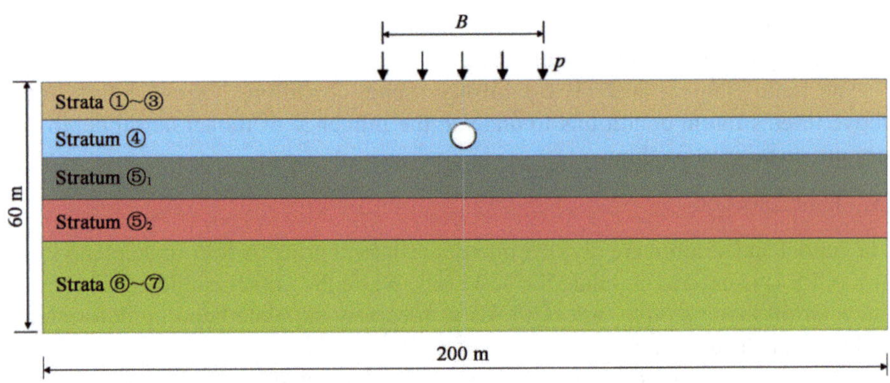

Fig. 3.6 Schematic of the finite element model

3.2.2 Parametric Analysis

(1) Load magnitude

A 0–100 kPa load is incrementally applied over a 20 m width at ground level to analyze the stress and deformation characteristics of tunnels under different stratum conditions (④–④, ④–⑤, and ⑤–⑤), with working condition settings detailed in Table 3.2.

Figure 3.7 shows the deformation of three tunnel types before and after load application (to better visualize deformation, the displacements in the figure have been magnified by a factor of 5). Under the load, the tunnel crown sags downward, and the haunches on both sides bulge outward, resulting in an overall "horizontal ellipse".

Table 3.2 Working condition settings for load magnitude effect'

	Serial number	Load value p/kPa	Load width *B*/m	Types of tunnel sections	Difference between soft and hard strata
Load magnitude	1	0–100 0–100	20	④–④ ④–④	Consistent with Table 3.1
	2			④–⑤ ④–5	
	3			⑤–⑤ ⑤–⑤	

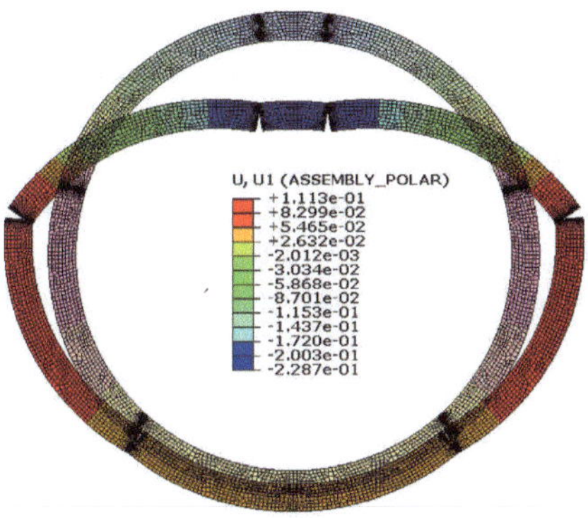

(a) Tunnel type ④④

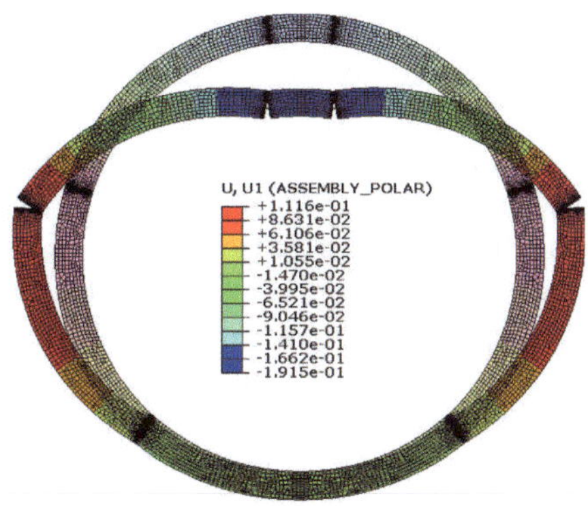

(b) Tunnel type④-⑤

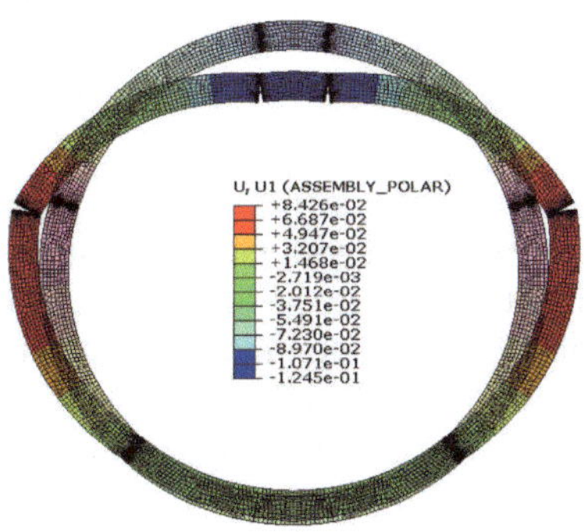

(c) Tunnel type ⑤- ⑤

Fig. 3.7 Deformation state of tunnel before and after load application

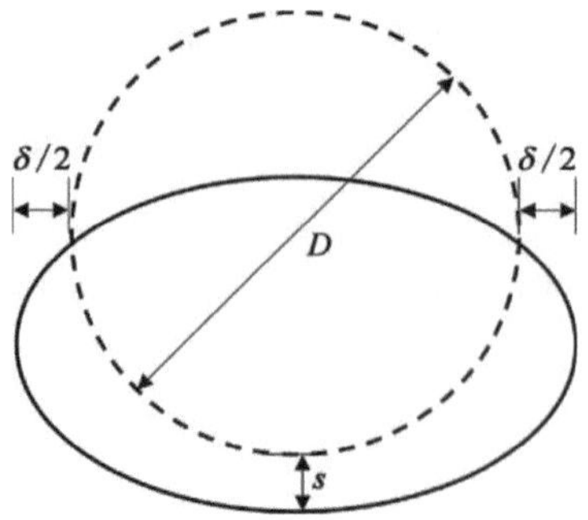

Fig. 3.8 Shield tunnel spatial configuration under load

The deformation of the lining segments is primarily caused by their bending and rotation around the joints. Under the load from above, the extrados of the top and bottom joints of the shield tunnel close while the intrados open, and the extrados of the shoulder joints open while the intrados close. This results in larger deformation in the top and shoulder joints, and smaller deformation in the bottom joints.

As shown in Fig. 3.7, under the load, the shield tunnel not only gets "flattened" but also experiences some settlement in the stratum. Therefore, the spatial configuration can be described in two aspects (Fig. 3.8): the transverse convergence deformation δ of the horizontal section and the settlement $\boldsymbol{s}$ at the invert.

The dimensionless index δ/D (i.e., the ratio of convergence deformation δ to tunnel outer diameter D) provides a more precise assessment of transverse deformation development in tunnels. Ultimate failure tests on tunnel segments revealed that:

When δ ranges between 2‰D and 5‰D, tunnels satisfy normal load-bearing capacity requirements under service conditions;

When δ reaches 20‰D–50‰D, tunnels approach their ultimate load-bearing limit state. Near the upper threshold ($\delta \approx$ 50‰D), structural collapse becomes imminent due to critical stress concentrations.

Figure 3.9 shows how δ/D changes with increasing load p for three tunnel types. At low loads, δ increases nearly linearly with load. When the load magnitude reaches a critical threshold, the stiffness of longitudinal joints and the nonlinear behavior of the surrounding soil mass emerge. Consequently, as load increases, the rate of horizontal convergence accelerates progressively, accompanied by a significant stiffness degradation of the tunnel structure.

At p = 20 kPa, the horizontal convergence δ values are 3‰D, 2.5‰D, and 1.7‰D for the three tunnel types, respectively, demonstrating satisfactory load-bearing capacity and confirming compliance with the 20 kPa threshold stipulated in relevant specifications. Exceeding specified thresholds of 30, 35, and 50 kPa, results in δ values surpassing or approaching 5‰D, necessitating timely unloading to mitigate significant transverse deformation. At p = 100 kPa, δ values reach 22‰D, 18‰D, and 12‰D. Notably, Tunnel ④–④ and Tunnel ④–⑤ approach their ultimate load-bearing capacity, while Tunnel ⑤–⑤ exhibits significant structural deformation, requiring immediate repairs to avert potential accidents.

Fig. 3.9 Development of horizontal convergence of the tunnel under loading

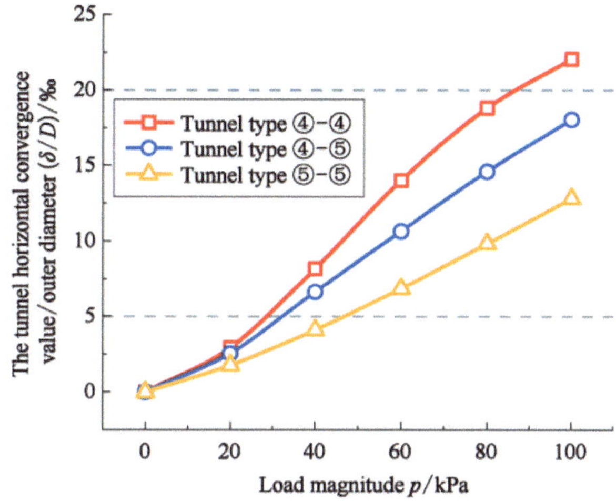

Figure 3.10 shows the development of tunnel invert settlement *s* under loading. The settlement values of tunnel invert exhibit a nonlinear development trend with increasing loading due to the nonlinear behavior of the surrounding soil mass. Under identical loading conditions, the settlement ratios for Type ④–④, ④–⑤, and ⑤–⑤ tunnels are approximately 1:0.52:0.22, indicating that deeper burial depths and stronger mechanical properties of the underlying strata correlate with reduced settlement. At a loading value of 100 kPa, the corresponding settlement values are 72, 40, and 17 mm for the three tunnel types, respectively.

Horizontal convergence and invert settlement serve as key indicators of tunnel spatial deformation under loading. As illustrated in Fig. 3.11, a strong linear correlation exists between these two parameters.

Fig. 3.10 Development of tunnel invert settlement under loading

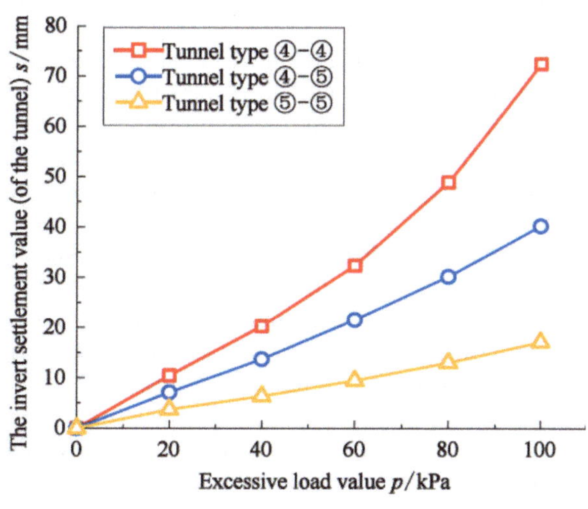

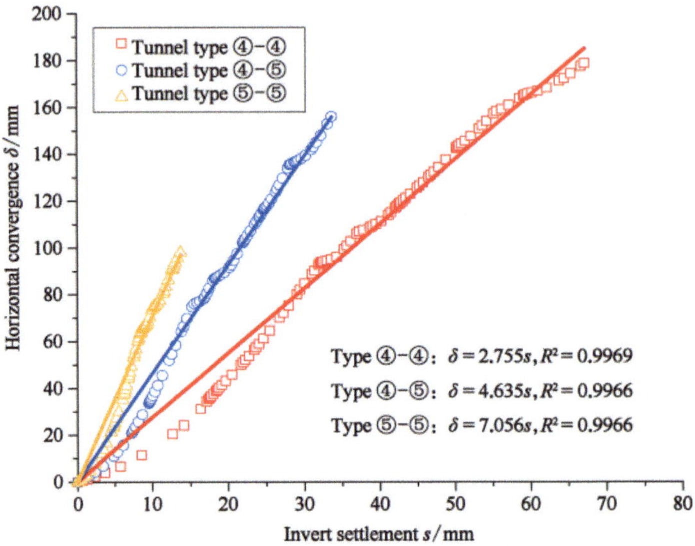

Fig. 3.11 Correlation between horizontal convergence and invert settlement

Figure 3.12 illustrates the stress-deformation behavior of the top and shoulder joints for loading values p of 0, 50, and 100 kPa, respectively. The stress cloud diagram presents the distribution of normal stress perpendicular to the cross-sectional plane. Positive stress values (indicated in red) signify tensile stress, while negative values (shown in blue) denote compressive stress.

When the loading value p is 0 kPa, no visible joint opening is observed. At this stage, the concrete at the bolted joint locations is primarily under tensile stress. Notably, there is a concentration of compressive stress in the concrete within the small areas on the outer side of the top joint (at 8°) and the inner side of the bottom joint (at 138°).

When p increases to 50 kPa, the joints exhibit noticeable opening. Specifically, at the top joint (8°), the extrados becomes closed while the intrados opens; conversely, at the shoulder joint (73°), the extrados opens while the intrados closes. The maximum tensile and compressive stresses on the concrete at the joint surfaces increase to varying degrees, though the locations of these maximum stresses remain relatively unchanged.

At a loading value of 100 kPa, the joints undergo more significant rotational deformation. At this point, the outer edge of the top joint (at 8°) becomes closed, and the inner edge of the shoulder joint (at 73°) also closes. Due to the participation of concrete in both the inner and outer edges of the joint in load-bearing, the concentration effect of compressive stress on the concrete is relatively reduced.

The development of joint rotation angles during incremental loading is analyzed for Type ④–④ tunnels, as shown in Fig. 3.13. With the increase of loading, the joint rotation development demonstrates a distinct nonlinear pattern. At 20 kPa loading, the

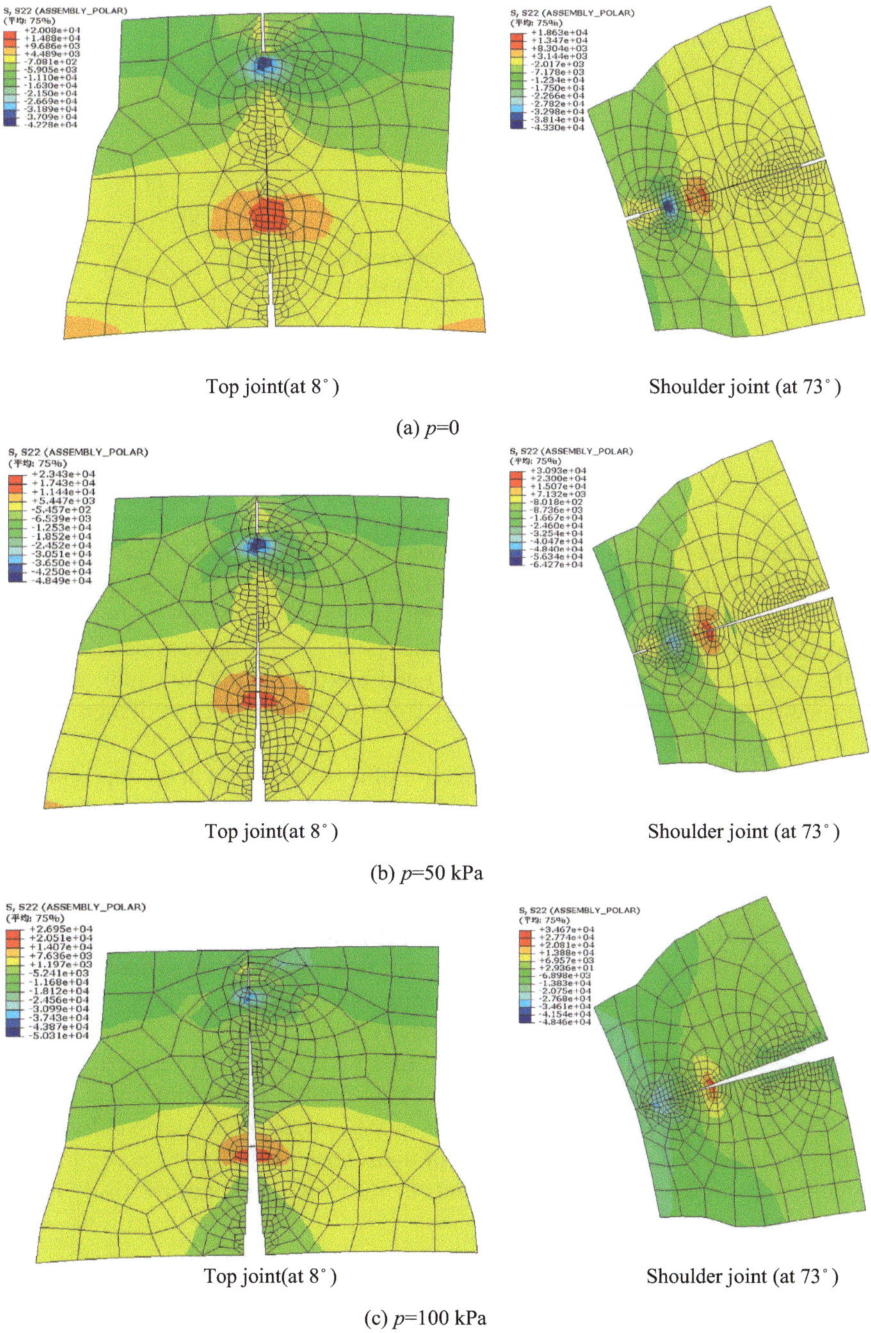

Fig. 3.12 Deformation of joints during incremental loading

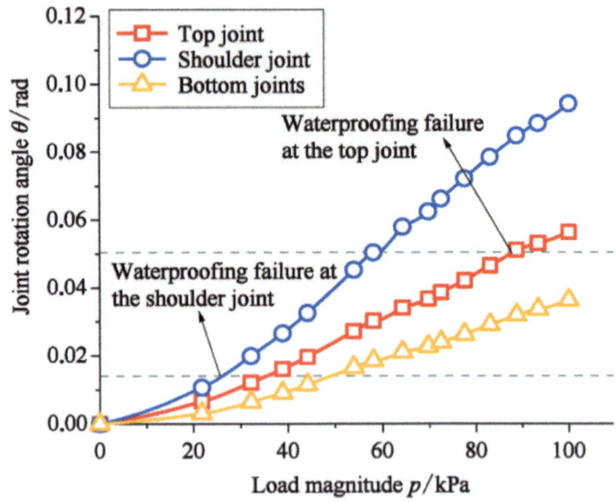

Fig. 3.13 Development of joint rotation angles during loading

rotation angles at the top joint (8°), shoulder joint (73°), and bottom joint (138°) are approximately 0.0066, 0.0107, and 0.0031 rad, respectively. These values increase significantly to 0.056, 0.094, and 0.036 rad at 100 kPa loading. In both loading conditions, the rotation angle at the shoulder joint (73°) is approximately the sum of the rotation angles at the top joint (8°) and bottom joint (138°), indicating that the transverse deformation of the tunnel structure is primarily attributed to joint rotational deformation.

For identical loading conditions, the ratio of rotation angles at the top, shoulder, and bottom joints is roughly 1:1.7:0.6, with the shoulder joint exhibiting the largest rotation angle and the bottom joint the smallest. According to Gong et al. (2019), to maintain the tunnel's waterproofing capacity, the rotation angle of joints under positive moment should not exceed 0.0505 rad, and those under negative moment should not exceed 0.0141 rad. Consequently, when the loading value exceeds 25 kPa, the shoulder joint's extrados opens excessively, making it difficult to maintain waterproofing capacity. Similarly, when the loading value exceeds 90 kPa, the top joint also fails to maintain its waterproofing capacity.

As shown in Fig. 3.14, there is a strong linear correlation between joint rotation angles and horizontal convergence values. This further demonstrates that segment rotation about joints is a primary cause of tunnel lateral convergence deformation. It also indicates that most deformation in the tunnel lining ring is accommodated at the joints or splices, while the segments themselves experience relatively limited deformation.

(2) Effect of load width

With the ground load value p maintained at 100 kPa, the loading width B is sequentially set to 10, 20, 30, 40, 50, 60, 80, and 100 m to investigate the effect of ground load width on the transverse deformation of three types of shield tunnels with different geological strata. The working conditions are presented in Table 3.3.

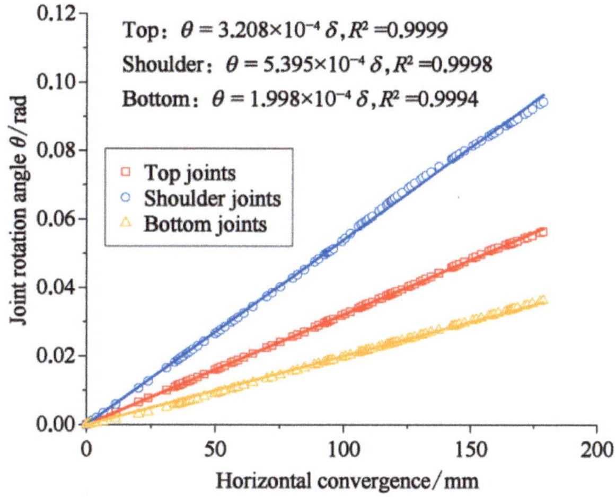

Fig. 3.14 Correlation between horizontal convergence and joint rotation angles

Table 3.3 Effect of load width on working conditions settings

	Load value p/kPa	Load width *B*/m	Types of tunnel sections	Difference between soft and hard strata
Load width	100	10, 20, 30, 40, 50, 60, 80, 100	④–④	Consistent with Table 3.1
			④–⑤	
			⑤–⑤	

Figure 3.15 presents the variation in horizontal convergence for three tunnel types under different load widths. When the load width is below 30 m (B/D = 5), tunnel horizontal convergence increases gradually with wider loading. When the load width exceeds 30 m (B/D = 5), additional increases in width do not yield greater transverse convergence.

For shallow-buried metro tunnels in soft soil, the impact of small-scale local loading above should also be given adequate attention. When the loading width is 10 m (*B/D* = 1.6), the horizontal convergence values of the ④–④, ④–⑤, and ⑤–⑤ type tunnels are 18‰*D*, 10‰*D*, and 3‰*D*, respectively. At this loading width, the transverse deformation of the ④–④ type tunnel is approaching the failure state.

Figure 3.16 presents the variation in invert settlement for the three tunnel types under different loading widths. As loading width increases, tunnel invert settlement initially rises sharply before stabilizing. In Fig. 3.15, when the overload width exceeds a certain threshold, tunnel horizontal convergence slightly decreases. In Fig. 3.16, when the overload width exceeds this threshold, tunnel invert settlement stabilizes. This occurs because, with larger overload widths, further increases in width result in relatively small additional vertical stress in the soil body, while horizontal stress continues to increase. Consequently, tunnel convergence decreases slightly, while settlement remains stable.

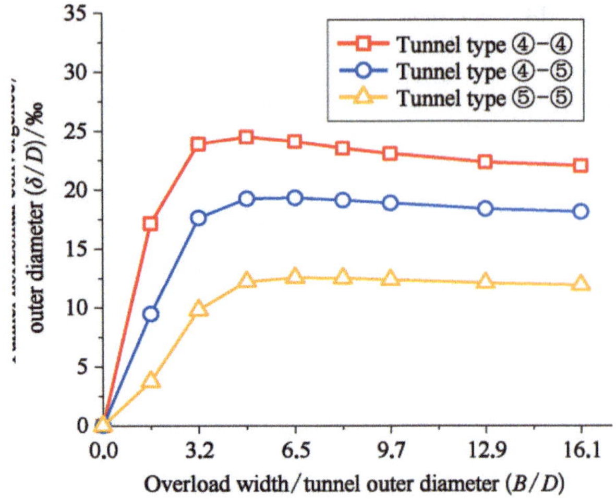

Fig. 3.15 Variation in horizontal convergence under different load widths

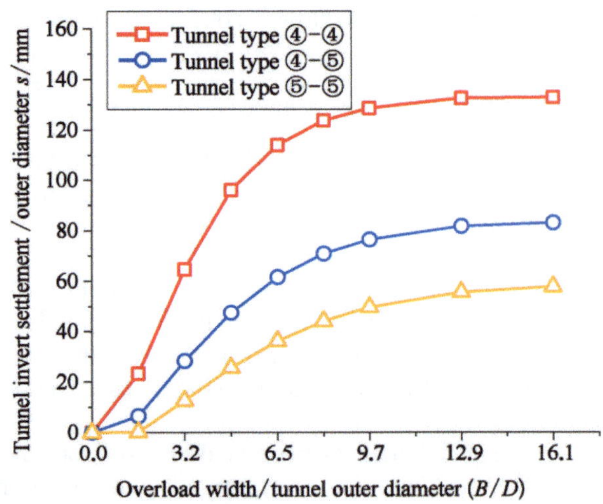

Fig. 3.16 Variation in invert settlement under different load widths

(3) Distance between the load and tunnel

The analysis was conducted with the loading value p fixed at 100 kPa and the load width B maintained at 20 m, while systematically varying the horizontal distance L between the outer edge of the load and the tunnel's outer contour line. The specific L values tested are −13.1, −3.1, 0, 3, 5, and 10 m. This experimental setup enables us to investigate how ground loading distance affects the transverse deformation of three types of shield tunnels with different surrounding strata. Table 3.4 details the working conditions, and Fig. 3.17 provides schematic illustrations of the setup.

Table 3.4 Effect of loading distance on working condition settings

	Loading value p/kPa	Loading distance L/m	Types of tunnel sections	Difference between soft and hard strata of the earth (geology)
Loading distance	100	−13.1, −3.1, 0, 3, 5, 10	④–④	Consistent with Table 3.1
			④–⑤	
			⑤–⑤	

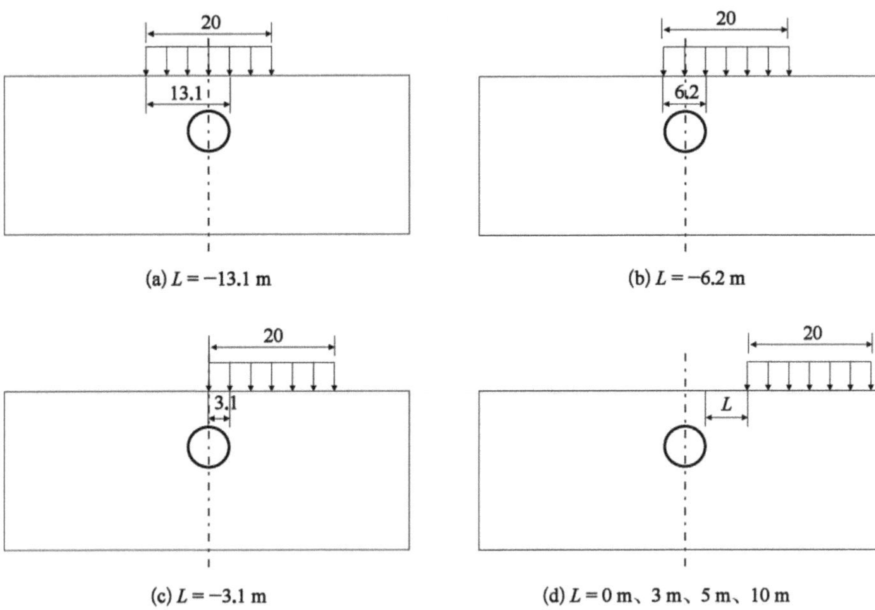

Fig. 3.17 Schematic illustrations of effect of loading distance on working conditions (unit: m)

Taking the ④–④ type tunnel for example, Fig. 3.18 shows the deformation of the tunnel under ground loading at different distances (with deformations magnified by 5 times for clarity). As shown in the figure:

① When the ground loading is directly above the tunnel, as shown in Fig. 3.18a, significant transverse deformation occurs. The tunnel assumes a "transverse elliptical" shape, with both sides of the haunch bulging outward. The top and shoulder longitudinal joints exhibit marked opening.

② As the distance between the ground loading and the tunnel increases, the load pressure direction shifts from directly above to obliquely above. Consequently, the tunnel displaces away from the loading direction. The arch waist on the side farther from the loading source displaces significantly more than that on the opposite side. Although some convergence deformation occurs, it is notably less

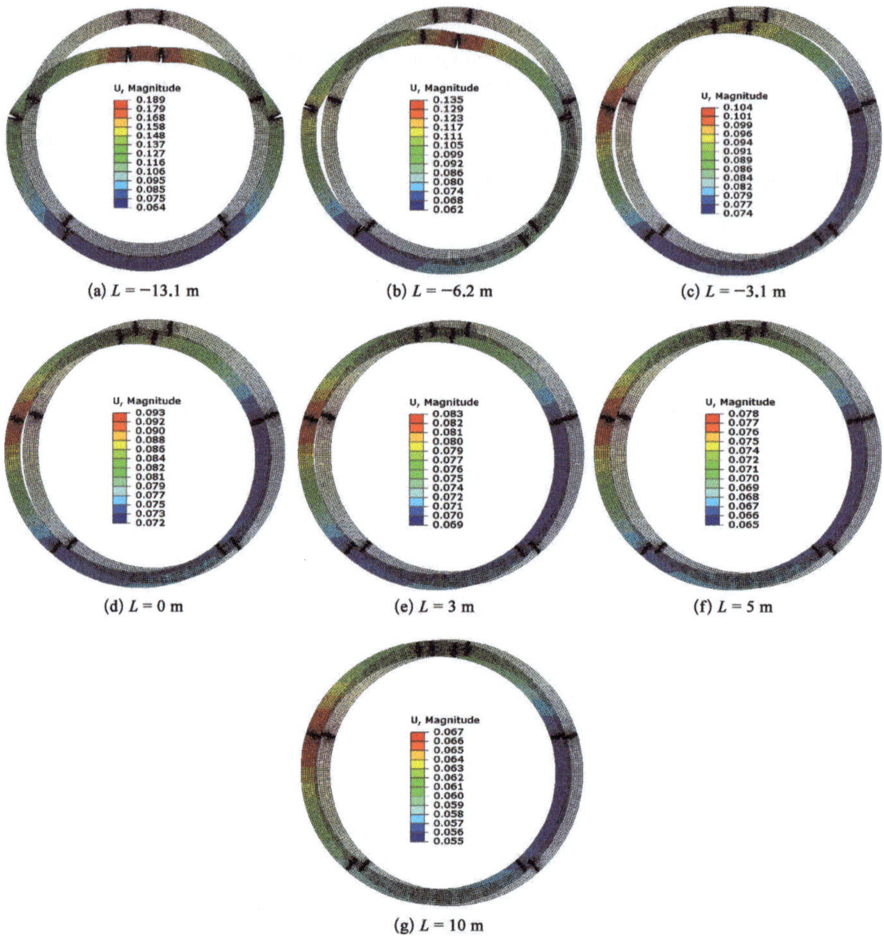

Fig. 3.18 Tunnel deformation under different loading distances

than when loaded directly above, and the tunnel takes on a “skewed elliptical” shape. At this point, the deformation of all joints is relatively minor.

The variation of horizontal convergence for the three tunnel types under the action of loading at different distances is shown in Fig. 3.19.

① As the loading distance increases from −13.1 m (directly above the tunnel) to −6.2 m, the horizontal convergence values of the ④–④, ④–⑤,and ⑤–⑤ type tunnels decrease from 22‰D, 18‰D, and 13‰D to 12‰D, 9‰D, and 8‰D, respectively. This suggests that for soft-soil shield tunnels, particularly shallow-buried ones, increasing the loading distance from the tunnel can significantly reduce adverse effects and protect the subway tunnel when ground loading is inevitable.

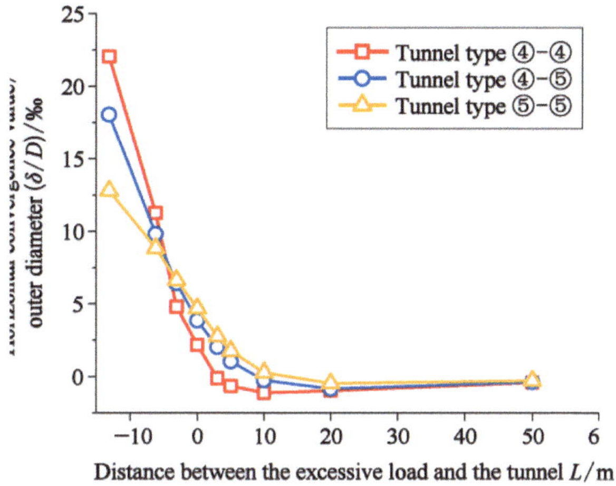

Fig. 3.19 Horizontal convergence values of the tunnel under different distance loading effects

② When the loading distance is 3 m, the convergence values of the three types of tunnels caused by 100 kPa loading within a 20-m range on the ground are all below 5‰*D*. At this point, the tunnel structure is relatively stable. This indicates that for subway tunnels in soft soil, the area within 3 m outside the tunnel's outer contour line is a sensitive zone. Engineering activities within this range should be carried out with great caution.

③ At a loading distance of 20 m (corresponding to the actual tunnel burial depth and located outside the main damage surface), loading-induced tunnel convergence deformation approaches zero.

Therefore, the area within the main damage surface of metro tunnels in soft soil is a deformation-sensitive zone, and large-scale engineering activities should be avoided outside this range as much as possible.

Analysis of Fig. 3.18 reveals that under eccentric loading, tunnels not only converge but also exhibit horizontal displacement away from the load. Figure 3.20 illustrates the horizontal displacement variation of tunnels under different loading distances. When $L < 0$ m (tunnel within the ground loading projection range), horizontal displacement increases rapidly with loading distance and then stabilizes. When $L > 0$ m (tunnel outside this range), horizontal displacement decreases as loading distance increases. Maximum displacement occurs at $L = 0$ m, indicating that loading adjacent to the tunnel's outer contour line is more likely to cause excessive horizontal displacement. In practical engineering, if such situations arise, assessments should focus not only on convergence and settlement but also on potential damage from excessive horizontal displacement.

Figure 3.21 presents the variation in tunnel invert settlement values under different loading distances. As the loading distance increases, the tunnel invert settlement value decreases.

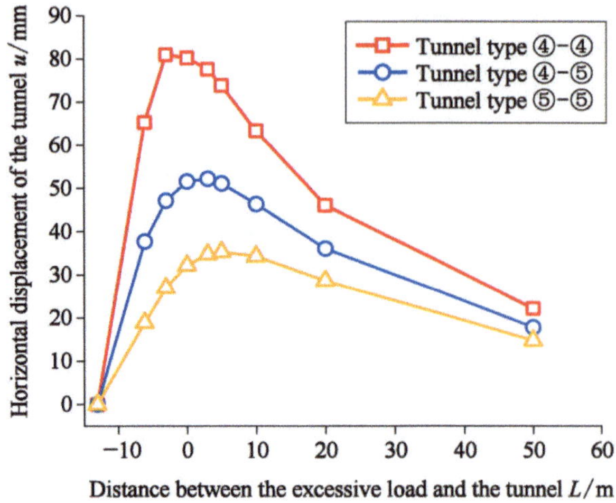

Fig. 3.20 Tunnel horizontal displacement under various loading distances

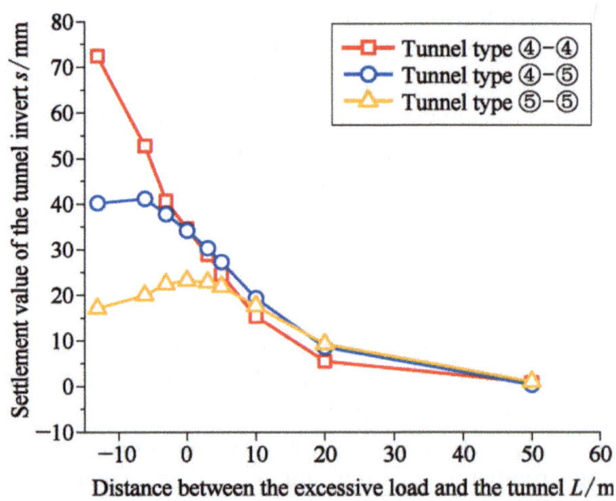

Fig. 3.21 Settlement value of the tunnel invert under different loading distance

Table 3.5 presents reference values for key deformation metrics of three tunnel types under varying load distances and magnitudes with a constant loading width B of 20 m.

(4) Stiffness of the Strata Tunnel Traversing

With the loading value p maintained at 100 kPa and the loading width B kept at 20 m, the modulus of elasticity of the soil strata ⑤$_1$, ⑤$_2$, and ⑥–⑦ (E_{lower}) is fixed at 150 MPa. Meanwhile, the modulus of elasticity of the soil stratum ④ (E_{upper}) is set to 10, 15, 30, 75, and 150 MPa in sequence to investigate its effect on tunnel deformation. Table 3.6 details the working conditions. To control variables and prevent the soil's internal friction angle and cohesion from affecting the calculation convergence and results, an elastic model is applied to the soil body in this working condition.

Table 3.5 Deformation indexes of shanghai shield tunnels with typical sections under ground loading

	Load value (kPa)	Loading distance/m								
		−13.1	−6.2	−3.1	0	3	5	10	20	50
④–④ tunnels										
Horizontal convergence value/(‰D)	20	2.92	1.28	0.92	0.47	0.05	−0.11	−0.27	−0.26	−0.17
	40	8.16	3.46	2.13	1.08	0.17	−0.12	−0.45	−0.42	−0.22
	60	14.00	5.90	3.24	1.48	0.20	−0.22	−0.65	−0.59	−0.28
	80	18.82	8.93	4.12	1.89	0.13	−0.40	−0.89	−0.77	−0.33
	100	22.06	11.28	4.81	2.19	−0.11	−0.63	−1.10	−0.97	−0.40
Horizontal displacement value/mm	20	0	5.32	6.92	8.01	8.29	8.18	7.34	5.23	2.62
	40	0	16.31	21.70	23.83	23.78	22.93	19.90	14.19	6.87
	60	0	32.93	39.70	41.25	40.36	38.71	33.41	24.01	11.55
	80	0	48.66	59.37	59.84	58.05	55.58	47.95	34.68	16.66
	100	0	65.36	81.00	80.22	77.62	73.90	63.38	46.13	22.16
Invert settlement value/mm	20	10.48	10.16	9.42	8.42	7.26	6.39	4.40	2.13	1.24
	40	20.28	18.55	17.61	15.81	13.49	11.63	7.54	3.11	1.16
Invert settlement value/mm	60	32.36	28.98	25.74	22.61	18.96	16.22	10.26	3.89	1.04
	80	48.93	40.44	33.21	28.79	24.04	20.54	12.87	4.69	0.93
	100	72.44	52.80	40.69	34.62	28.88	24.58	15.31	5.49	0.83
Tunnel ④–⑤										
Horizontal convergence value/(‰D)	20	2.54	2.05	1.56	1.11	0.74	0.48	0.11	−0.11	−0.05
	40	6.62	4.21	2.99	2.02	1.24	0.76	0.11	−0.25	−0.12
	60	10.64	6.48	4.23	2.68	1.56	0.90	−0.01	−0.45	−0.20
	80	14.59	7.96	5.36	3.30	1.80	0.99	−0.14	−0.64	−0.28
	100	18.05	9.83	6.41	3.86	2.03	1.05	−0.25	−0.83	−0.36
Horizontal displacement value/mm	20	0	3.63	4.64	5.36	5.69	5.71	5.51	4.44	2.13
	40	0	9.24	12.04	13.97	14.80	14.78	13.87	11.10	5.54
	60	0	16.85	22.73	25.89	26.72	26.35	24.14	18.90	9.31
	80	0	27.10	34.88	38.49	39.09	38.34	34.91	27.28	13.41
	100	0	37.71	47.16	51.59	52.21	51.12	46.38	36.07	17.79
Invert settlement value/mm	20	7.13	7.52	6.94	6.32	5.73	5.25	3.83	1.57	−0.02
	40	13.72	14.48	14.24	13.33	12.13	11.00	7.86	3.37	0.06
	60	21.54	22.11	22.41	20.80	18.53	16.65	11.78	5.08	0.11
	80	30.17	31.41	30.34	27.67	24.54	22.02	15.59	6.81	0.19
	100	40.20	41.14	37.78	34.19	30.33	27.25	19.34	8.55	0.31

(continued)

Table 3.5 (continued)

	Load value (kPa)	Loading distance/m								
		−13.1	−6.2	−3.1	0	3	5	10	20	50
Tunnel ⑤–⑤										
Horizontal convergence value/(‰D)	20	1.78	1.95	1.66	1.24	0.85	0.61	0.15	−0.04	−0.05
	40	4.10	4.03	3.29	2.39	1.56	1.06	0.25	−0.10	−0.09
	60	6.84	5.84	4.93	3.42	2.09	1.36	0.29	−0.19	−0.15
	80	9.82	7.39	5.84	4.21	2.44	1.56	0.29	−0.30	−0.20
	100	12.81	8.86	6.67	4.70	2.77	1.76	0.29	−0.45	−0.27
Horizontal displacement value/mm	20	0	2.40	3.40	4.02	4.34	4.41	4.32	3.71	1.82
	40	0	5.59	7.65	9.08	9.89	10.15	10.11	8.79	4.64
	60	0	9.18	12.46	15.15	16.86	17.41	17.37	14.92	7.79
	80	0	13.13	18.93	23.10	25.41	26.10	25.68	21.58	11.18
	100	0	18.91	27.00	32.21	34.79	35.33	34.32	28.60	14.77
Invert settlement value/mm	20	3.76	4.01	4.17	4.17	4.02	3.85	3.22	1.62	0.09
	40	6.39	7.40	8.01	8.33	8.26	8.04	6.71	3.45	0.30
	60	9.48	11.27	12.04	12.86	12.93	12.57	10.30	5.32	0.49
	80	13.05	15.40	17.16	17.92	17.94	17.29	14.00	7.25	0.72
	100	17.14	20.03	22.48	23.22	22.86	21.90	17.67	9.24	0.99

Table 3.6 Effect of stiffness of the strata tunnel traversing on working conditions settings

	Load value p/kPa	Load width *B*/*m*	Type of cross-section of the tunnel	Stratum elasticity tunnel traversing modulus E_{upper}/MPa	Elasticity of underlying stratum modulus E_{lower}/MPa
Difference between the hard and soft strata	100	20	④–④	10, 15, 30, 75, 150	150

Figure 3.22 presents the tunnel's horizontal convergence and the variation of its invert settlement with changes in the elastic modulus of the stratum tunnel traverses.

① When the tunnel's underlying stratum is stiff (with a modulus of elasticity maintained at 150 MPa in this scenario), the tunnel invert settlement under loading remains relatively low, decreasing only slightly as the elastic modulus of the traversing stratum increases.

② As the elastic modulus of the stratum traversed by the tunnel increases, the horizontal convergence of the tunnel rapidly declines.

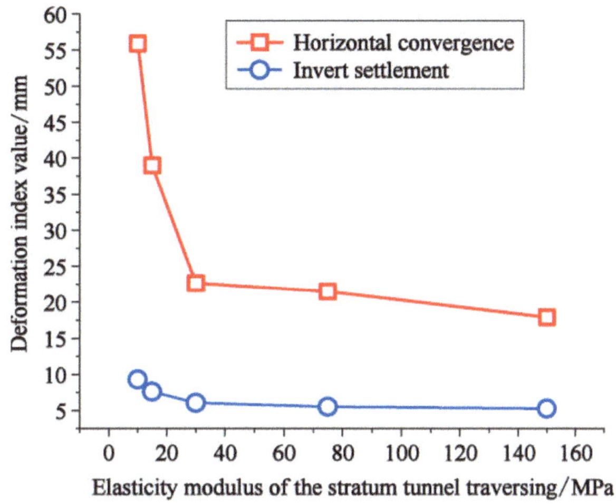

Fig. 3.22 Variation of deformation index with changes in stiffness of strata tunnel traversing

(5) Stiffness of the tunnel's underlying stratum

With the loading value *p* maintained at 100 kPa, the loading width B kept at 20 m, and the elastic modulus of Stratum ④ (E_{upper}) fixed at 10 MPa, the elastic modulus of strata ⑤$_1$, ⑤$_2$, and ⑥–⑦ (E_{lower}) varies sequentially to 10, 20, 50, 100, and 150 MPa. This approach is employed to investigate the effect of the stiffness of the underlying stratum on tunnel deformation. The specific working conditions are detailed in Table 3.7. In this set of working conditions, the soil body is modeled using an elastic model.

Figure 3.23 presents the variation of the tunnel's horizontal convergence and invert settlement with the changes of underlying stratum's elastic modulus. The figure shows that:

① When the tunnel is in soft soil (with the traversing stratum's elastic modulus kept at 10 MPa), the tunnel's horizontal convergence decreases only slightly as the underlying stratum's elastic modulus increases.

② As the underlying stratum's elastic modulus increases, the tunnel's invert settlement decreases rapidly.

Table 3.7 Effect of elasticity modulus of underlying stratum on working condition setting

	Load value p/kPa	Load width *B*/*m*	Type of cross-section of tunnel	Elasticity modulus of stratum tunnel traversing *Eupper*/MPa	Elasticity modulus of the underlying stratum tunnel traversing *Elower*/MPa
Difference in stiffness of strata	100	20	④–④	10	10, 20, 50, 100, 150

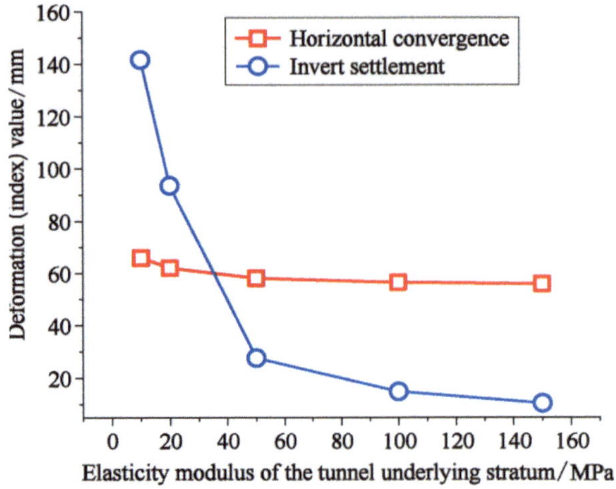

Fig. 3.23 Variation of deformation (index) with changes in stiffness of the tunnel underlying strata

Combining the discussions in (4) and (5) on the effect of the tunnel's traversing stratum and underlying stratum's stiffness on its horizontal convergence and invert settlement under ground loading, the following conclusions can be drawn:

① The tunnel's horizontal convergence under ground loading mainly depends on the traversing stratum's soil properties. A softer traversing layer makes the tunnel more prone to excessive convergence deformation, while a harder layer reduces this deformation under loading.

② The tunnel's invert settlement under ground loading is primarily influenced by the underlying stratum's soil properties. A softer underlying layer increases the risk of excessive settlement, while a harder layer reduces it.

For tunnels in soft soil strata, a soft underlying layer makes structural damage due to excessive settlement more likely. For tunnels constructed in soft soil strata, the presence of a softer underlying lying stratum increases the likelihood of structural damage resulting from excessive settlement. As the underlying stratum's elastic modulus increases, the risk of excessive settlement decreases, but the risk of excessive convergence rises. With a hard underlying stratum, the tunnel is less likely to be damaged by excessive settlement but more prone to structural damage from excessive convergence. It is essential to effectively manage the deformation characteristics of soft-soil shield tunnels and develop customized monitoring and protection schemes for tunnels constructed in varying stratigraphic conditions.

3.2.3 Summary of Deformation Patterns

In the previous section, ABAQUS finite element software is applied to analyze the effect of loading magnitude, width, distance, tunnel burial depth, stratum properties, and other factors on the deformation behavior of tunnels under ground loading. Through parametric analysis, the following key deformation patterns are identified.

① The transverse deformation of shield tunnels under ground loading primarily manifests as transverse convergence and invert settlement. Under eccentric loading, horizontal displacement also occurs. Transverse convergence reflects the bending and rotational deformation of tunnel segments around joints, while invert settlement and horizontal displacement indicate the tunnel's overall movement within the stratum. As the loading value p increases, tunnel convergence, invert settlement, horizontal displacement, and joint rotation all exhibit nonlinear growth.

② To ensure the structural safety of the tunnel and meet its waterproofing requirements, the ground loading value should be limited to no more than 20 kPa when the tunnel is in a good initial condition.

③ When the loading width is less than 30 m (5 times the tunnel diameter), the tunnel's transverse convergence increases with loading width. When the loading width exceeds 30 m, further increases result in slight decreases in transverse convergence. Additionally, the effect of small-scale local loading above the tunnel on its transverse deformation should be given adequate attention.

④ As the distance of ground loading from the tunnel increases, the tunnel's horizontal convergence and invert settlement gradually decrease, but some horizontal displacement occurs in the direction away from the loading. Sensitive areas are located within the tunnel's main damage surface, and engineering activities should be carried out with caution within these areas.

⑤ The horizontal convergence of a tunnel under ground loading is primarily influenced by the soil properties of the strata traversed by the tunnel, while the invert settlement is mainly affected by the soil properties of the underlying strata. For soft soil tunnels, when the underlying stratum is soft, emphasis should be placed on monitoring settlement deformation; when the underlying stratum is hard, attention should be focused on convergence deformation. Monitoring and protection measures should be implemented in a targeted manner accordingly.

3.3 Model Test Analysis

3.3.1 Introduction to the Principles of Model Testing

Testing methods play a vital role in civil engineering research. Tests on the mechanical properties of shield tunnels are primarily categorized into prototype tests and model tests. Prototype tests, conducted on-site, provide authentic first-hand data but are limited by their large scale, high cost, extended duration, and susceptibility to interference. In contrast, model tests, which involve scaled-down representations of prototypes, offer a targeted and cost-effective means of evaluating tunnel structural performance in a controlled indoor environment.

Model tests emulate prototype structures based on scaling similarity principles. A comprehensive understanding of the similarity between the model and the prototype is essential. Typically, model test results are scaled up by a similarity constant to predict the corresponding characteristics of the prototype. However, this scaling process also magnifies any errors inherent in the model tests. Excessively high similarity constants may distort the model test outcomes, while lower constants increase the demands on the test site and equipment. Therefore, the formulation of model-test similarity relationships must reconcile economic constraints with reliability requirements.

Constrained by various conditions, prototype tests demand substantial human and material resources, and are time-consuming, labor-intensive, and prone to errors. In contrast, model tests are not only economical and data-reliable but also easy to operate and expedite the process, making them the preferred experimental approach for researchers. To draw conclusions consistent with those of prototype tests, the similarity triple theorem must be strictly followed during the simplification process to achieve the desired outcomes.

The similarity triple theorem provides the theoretical basis for model tests. It not only promotes the development of similarity theory but also allows for focusing on primary factors and ignoring secondary ones based on their influence on model test phenomena. This approach yields approximate prototype test results.

By applying the three theorems of similarity, prototype tests can be simplified into model tests that are easy to operate, reliable in test data, and feasible in technical solutions. This simplification is achieved through geometric similarity, physical similarity, similarity in initial conditions, and similarity in boundary conditions. Prototype tests are not only time-consuming and labor-intensive but also challenging to implement. However, as long as the purpose of the test is clear and the test is well-prepared, the three theorems of similarity can be applied in a scientific and rational manner to derive accurate conclusions.

As an assembled structure of segments, subway shield tunnels are significantly affected in their mechanical properties by the ring and longitudinal joints. How to achieve reasonable similarity for these joints in model tests remains one of the key challenges.

3.3.2 Model Test Design

(1) Similarity ratio design

The similarity design of indoor model tests generally begins with the determination of similarity constants for fundamental physical quantities such as geometry, gravity, and elastic modulus. Subsequently, the similarity ratios of the remaining parameters are established based on the principles of similarity or dimensional analysis (Huang et al. 2019).

In this experiment, the model shield tunnel was designed using a geometric similarity constant (C_l) = 10, a gravitational acceleration similarity constant (C_g) = 1, and an elastic modulus similarity constant (C_E) = 10. The similarity relationships for various physical quantities are presented in Table 3.8 (Fang 2022).

(2) Model tunnel design

The prototype shield tunnel, depicted in Fig. 3.24, is a subway shield tunnel constructed using through-seam assembly. It has an outer diameter of 8000 mm and a segment thickness of 300 mm.

The tunnel segments have a width of 2000 mm and are divided into six individual pieces. They are cast from C50 concrete, which has an elastic modulus of 30 GPa. At

Table 3.8 Similar constants

	Similarity relationship	Similar constant	Physical quantity	Similarity relationship	Similar constant
Geometric dimensions	Basic quantity	10	Bending moment	$C_M = C_\gamma C_l^4$	10 000
Capacity	Basic quantity	1	Axial forces	$C_N = C_\gamma C_l^3$	1 000
Strains	C_ε	1	Shearing force	$C_Q = C_\gamma C_l^3$	1 000
Stresses	$C_\sigma = C_l C_\gamma$	10	Bending stiffness	$C_{EI} = C_l^4$	10 000
Displacement	$C_\delta = C_l$	10	Axial stiffness	$C_{EA} = C_l^3$	1 000
Modulus of elasticity	$C_E C_l$	10	Shearing rigidity	$C_{GA} = C_l^3$	1 000

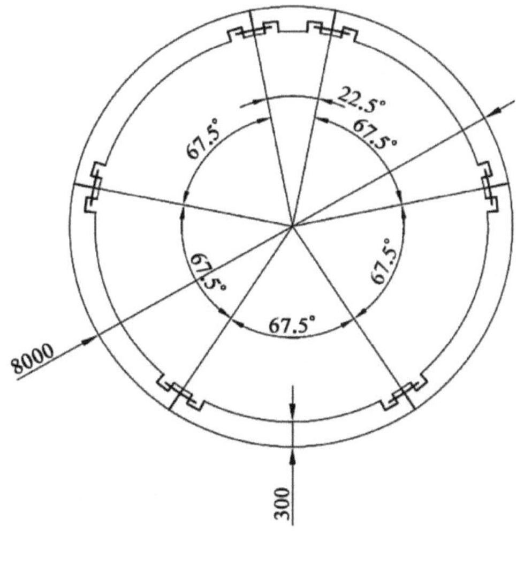

Fig. 3.24 Cross-section of the prototype shield tunnel (unit: mm)

the joints between segments, there are six longitudinal joints in total. Each of these joints is connected by two M48 circumferential bolts, which have an elastic modulus of 206 GPa.

Based on the geometric similarity ratio, the modeled shield tunnel has an outer diameter of 800 mm, a thickness of 30 mm, an inner diameter of 740 mm, and a width of 200 mm. The number and position of the lining segments in the model are consistent with the prototype. Specifically, the segments consist of one crown segment with a central angle of 22.5°and five standard segments each with a central angle of 67.5°.

The tunnel segments are fabricated from acrylic glass with a modulus of elasticity of 3 GPa, which is widely used in various structural modeling tests and can better reflect the mechanical behavior of tunnel segments under elastic conditions.

Rubber components are installed to simulate the longitudinal joints, and six segments are assembled into a full ring using model bolts. The design of the longitudinal joints is crucial. After extensive comparative testing, a rubber sheet measuring 200 mm × 30 mm × 5 mm with a Shore hardness of 55 is selected for simulating the joints.

Bolt equivalent design: With an elastic modulus of 206 GPa and a diameter of 48 mm, the prototype bolt's parameters are referenced to calculate the model bolt's section diameter, which is made of 304 alloy steel with an elastic modulus of 194 GPa. Following the tensile stiffness similarity principle (Yuan et al. 2020, 2017), the required diameter is 4.9 mm. Thus, two M5 bolts are used in the test for tunnel segments assembly.

As shown in Fig. 3.25, each end of the lining segments has reserved bolt holes and hand holes for bolt installation. A torque wrench is used to apply a preload force of 2.5 kN to the bolts.

The completed model is shown in Fig. 3.26.

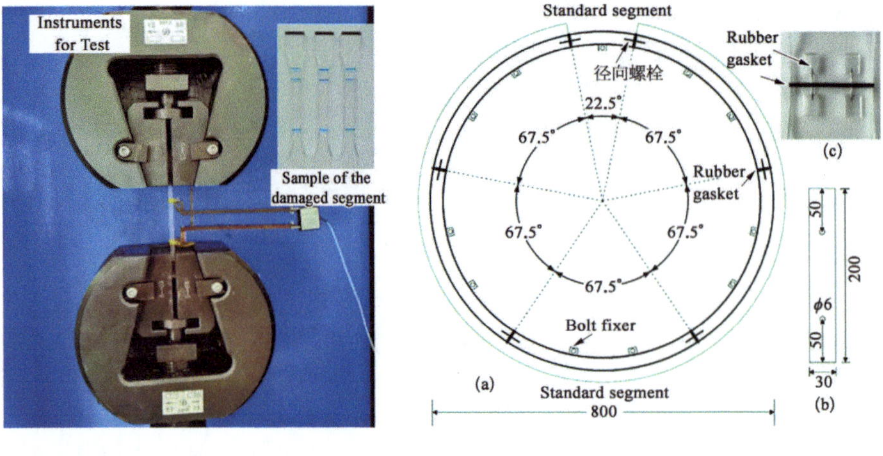

Fig. 3.25 Acrylic glass testing and processing (unit: mm)

Fig. 3.26 Model tunnel

(3) Model soil preparation

Huang et al. (2018) have pointed out that in elastic range model tests, the model soil's compressive modulus must satisfy similarity relationships, while cohesion and internal friction angle need not. Since this test focuses on tunnel deformation under ground-level localized loading, the unit weight of soil doesn't require strict similarity. Consequently, two model soils are formulated: sandy soil (fine sand) and sawdust soil (fine sand + sawdust), simulating different soil conditions. Their densities are 1400 and 860 kg/m^3, with internal friction angles of 35°and 30°, respectively. The soil compressive modulus is based on values corresponding to 10–20 kPa pressure. In the model, the compressive moduli of fine sand and sawdust soil are 2.65 and 0.85 MPa, respectively, while the corresponding prototype soil compressive moduli are 26.5 and 8.5 MPa, corresponding to sandy and soft soil strata, respectively.

(4) Sensor design

During the test, the convergence deformation of the lining segments, the opening of the lining joints, and the additional earth pressure on the outer surface of the tunnel are monitored. Convergence deformation, calculated by measuring lining diameter displacement, is monitored using four POM-HDQ20-50A inductive displacement gauges (with a 50 mm range and 0.5% accuracy) arranged at 0°, 45°, 90°, and 135° sections, as shown in Fig. 3.27a. The joint opening angle is calculated by measuring the deformation of the inner and outer surfaces of each lining joint, where each joint is equipped with a strain gauge on both surfaces, as depicted in Fig. 3.27b. Additional soil pressure is monitored using seven BWM-type earth pressure sensors installed at different angles on the outer surface of the lining, with a range of 0.02–20 MPa and an accuracy class of 0.2, as shown in Fig. 3.27c.The physical arrangement of these sensors is illustrated in Fig. 3.28.

To verify the reliability of the test, the aforementioned sensors are installed on two adjacent lining segments (A and B) subjected to the load. Figure 3.29a–c illustrate the convergence of the tunnel diameter at sections A and B under three working conditions (e = 0 mm, 200 mm, and 1000 mm) when the tunnel passes through

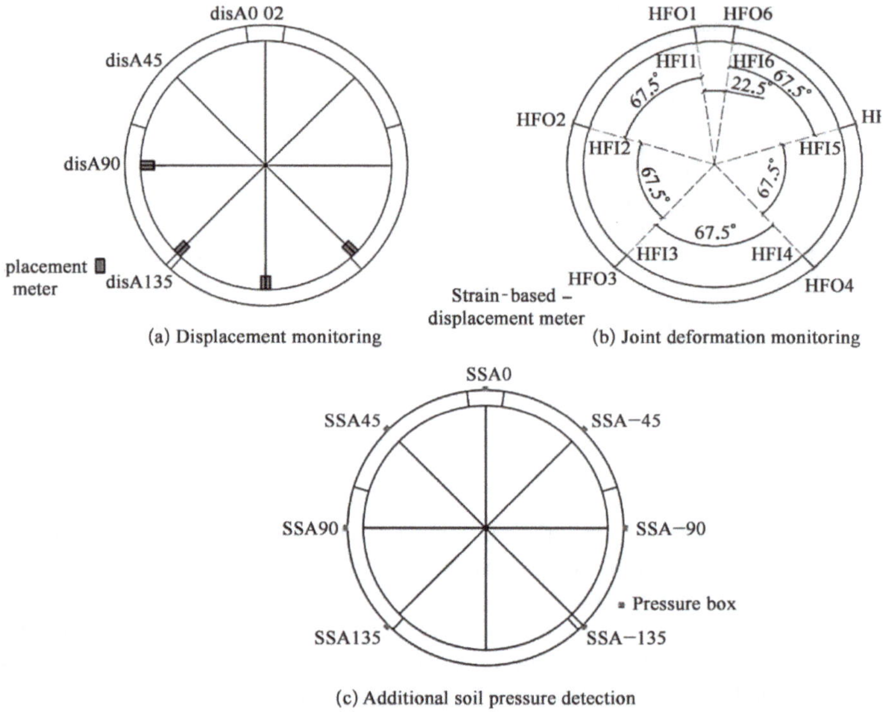

Fig. 3.27 Sensor arrangement

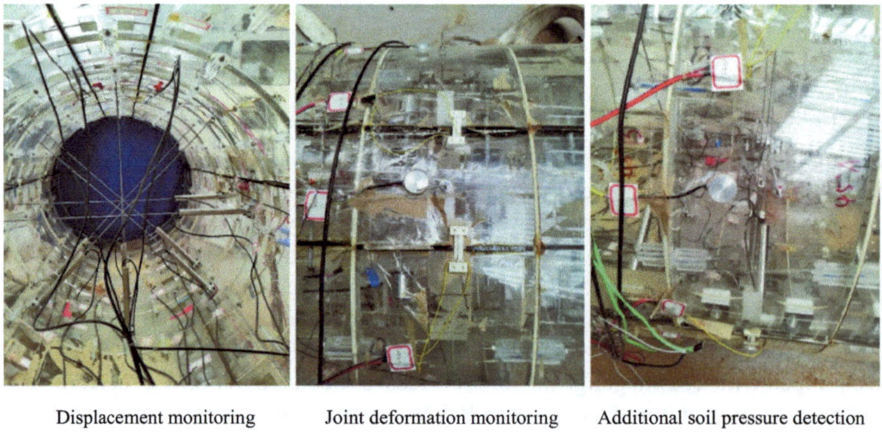

Fig. 3.28 Sensor arrangement physical diagram

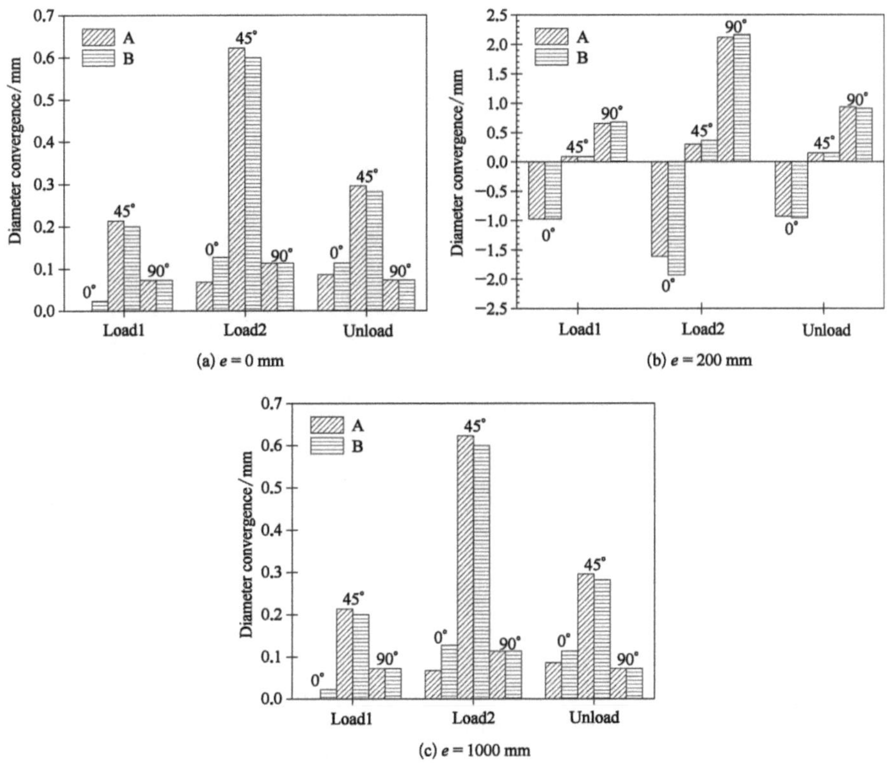

Fig. 3.29 Comparison of deformation between A and B segment rings

sandy soil. These figures show an average difference of only 0.08 mm in deformation across the same sections of the two lining rings, indicating the reliability of the test molding and testing methods. When processing the data, the average value of the corresponding positions of the two lining rings is used as the representative value for the test. Additionally, all measured deformations and forces in the test are adjusted by deducting the initial values recorded after the backfill deformation stabilizes.

3.3.3 Experimental Procedure

The test is conducted in the large soil box at the State Key Laboratory of Civil Engineering Disaster Prevention, Jiading Campus, Tongji University. This soil box measures 4 m in length, 3 m in width, and 2 m in height. As depicted in Fig. 3.30, the outer structure of this soil box is fabricated from steel sections, while its inner wall is lined with a rubber membrane.

In the test, a layer of fine sand with a thickness of $0.5Dt$ (where Dt is the tunnel outer diameter) was first laid as the underlying stratum for the tunnel. Then, the

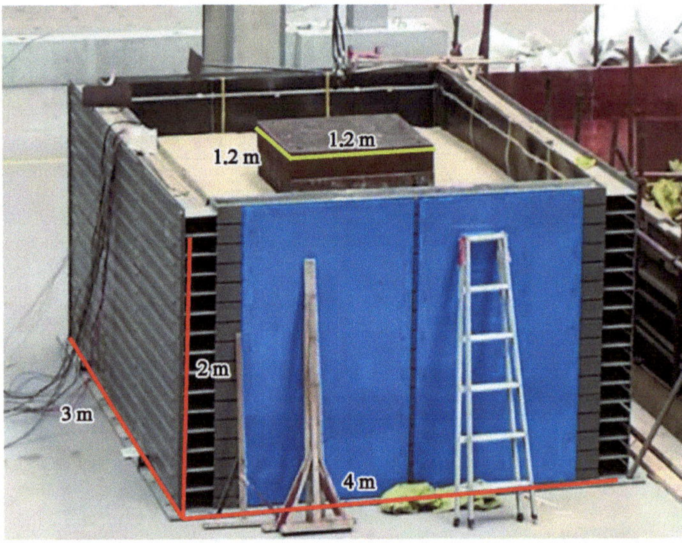

Fig. 3.30 Physical drawing of the model (state key laboratory of civil engineering disaster prevention)

tunnel model is lifted into the model tank. Using quality control methods (Huang et al. 2019), soil layers are filled sequentially, with each layer being 0.25Dt thick. Each layer is compacted as required and left for 24 h after placement to allow deformation to stabilize. After filling to the preset height, iron blocks are placed at predetermined locations to simulate localized ground loading. Each test is divided into two loading stages, with each stage involving 2 tons of iron blocks measuring 1.2 m × 1.2 m at the base, applying a pressure of $p = 14$ kPa per stage, for a total loading pressure of $p = 28$ kPa–conditions that simulate the common scenario of a sudden 2-m-high soil pile. After the first loading stage, the model is left to rest for half an hour before the second loading stage, after which it rests again for 0.5 h before unloading (Fang 2022).

To investigate the effect of the tunnel's surrounding stratum conditions, two model tests with different soil distributions are designed as shown in Fig. 3.31. These tests simulate tunnels passing through sandy soil (Stratum 1) and soft soil (Stratum 2).

To investigate the impact of local loading position on shield tunnels, three loading forms with eccentricity distances of $e = 0$ mm, $e = 200$ mm, and $e = 1000$ mm from the tunnel center-line are set up, as shown in Fig. 3.32. The detailed test conditions are listed in Table 3.9, and some photographs from the testing process are provided in Fig. 3.33.

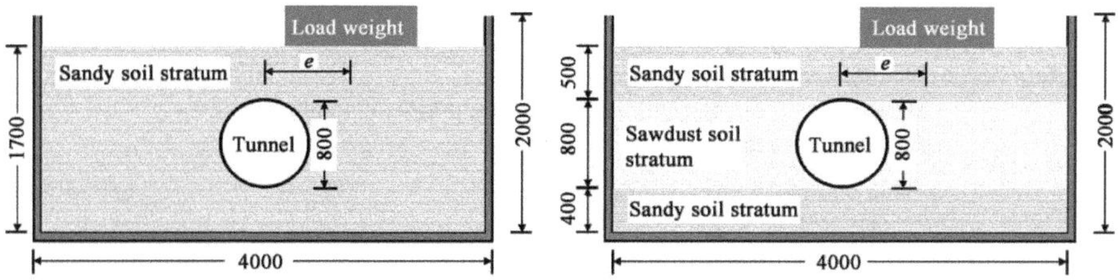

(a) Sandy soil stratum tunnel traversing(stratum 1)

(b) Soft soil stratum tunnel traversing(stratum 2)

Fig. 3.31 Conditions of soil tunnel traversing (in mm)

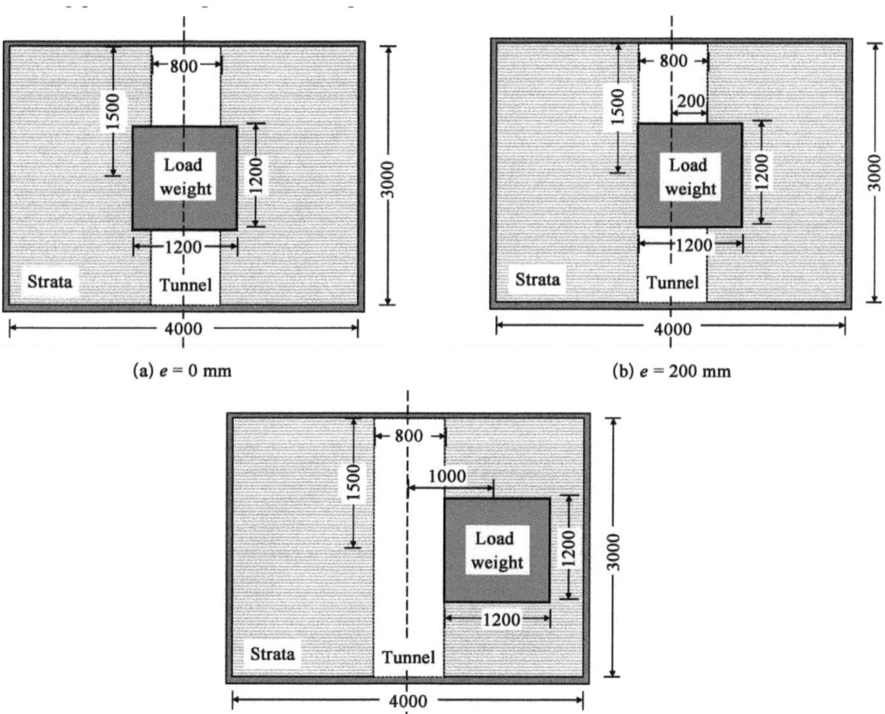

(a) $e = 0$ mm

(b) $e = 200$ mm

(c) $e = 1000$ mm

Fig. 3.32 Loading eccentricity conditions (unit: mm)

3.3.4 Analysis of Test Results

(1) Tunnel convergence deformation

Table 3.10 presents the deformation of each cross-section of the tunnel lining segments at $p = 28$ kPa. Positive deformation values indicate outward expansion of the monitoring cross-section, while negative values denote inward contraction.

Table 3.9 Working conditions of the test

Test number	Tunnel traversing stratum	Loading eccentricity e/mm	Primary loading p/kPa	Secondary loading p/kPa	Unloading
①	Sandy soil	0	14	28	Unloading
②		200			
③		1000			
④	Soft soil	0			
⑤		200			
⑥		1000			

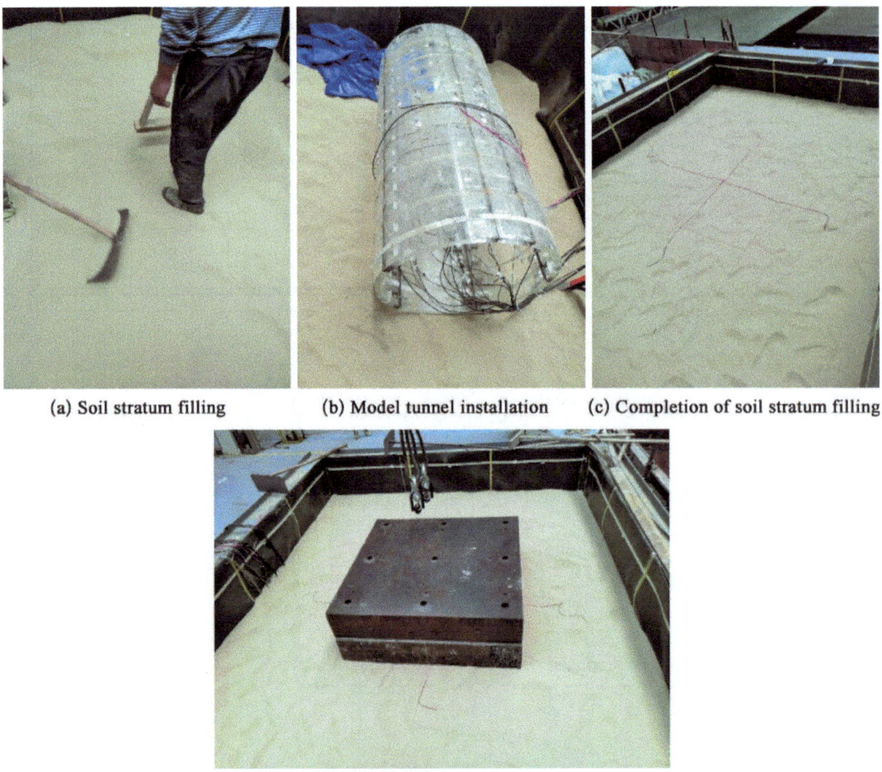

(a) Soil stratum filling (b) Model tunnel installation (c) Completion of soil stratum filling

(d) Load application

Fig. 3.33 Test procedure

As shown in the table, when e = 0 mm, the tunnel's vertical cross-section contracts inward, and the horizontal cross-section expands outward, with the contraction and expansion being roughly equivalent. The deformation of the inclined cross-section is relatively minor, resulting in transverse elliptical deformation of the lining segments. Under these conditions, the ratios of horizontal convergence to outer diameter (*ΔD/Dt*) for sand and soft soil tunnels are 6.76 and 15.98‰, respectively, both exceeding the normal value of 5.1‰ reported in the literature (Huang et al. 2017).

Based on the data in Table 3.10, the variations in horizontal convergence values of tunnel lining segments under different loading positions and stratigraphic conditions can be analyzed, as illustrated in Fig. 3.34. As the loading eccentricity increases, the horizontal convergence of the tunnel initially decreases rapidly and then gradually stabilizes. At $p = 28$ kPa, when e increases from 0 to 200 mm, the horizontal convergence values for tunnels in sandy soil layers and soft soil layers decrease by 60 and 73‰, respectively.

Furthermore, when e is 0, 200, and 1000 mm, the long-axis direction of the tunnel deformation is 90° (horizontal), 90°, and 45° (oblique), respectively. This indicates that increasing the loading eccentricity causes the tunnel lining to transition from transverse elliptical deformation to oblique elliptical deformation. In practical engineering, monitoring and reinforcement measures should be adjusted accordingly.

Comparison of different stratum properties shows that soft soil tunnels generally have larger horizontal convergence than sandy soil tunnels, with the maximum horizontal convergence being 2.5 times higher. This is because soft soil provide less resistance, making tunnel structures more prone to deformation under surface loading (Huang et al. 2019). Moreover, when $p = 28$ kPa, increasing e from 0 to 200 mm and 1000 mm results in a decrease in horizontal convergence of sandy soil tunnels by 60 and 90% respectively compared to when $e = 0$ mm. For soft soil tunnels, the decrease is 73 and 95% respectively. This indicates that for subway tunnels in soft soil, increasing the distance between the load and the tunnel can better protect the tunnel.

Analysis of the loading–unloading process reveals significant residual deformation of tunnel lining segments post-unloading. This indicates the adverse effects of local loading on tunnels are not fully reversible. Tunnel cross-sectional deformation typically originates from two aspects: deformation of the lining segment structure itself and dislocation deformation at the joints (Yi et al. 2019). The additional load from loading causes not only plastic deformation in soil strata but also dislocation

Table 3.10 Convergence deformation value of tunnel pipe diameter at $p = 28$ kPa (unit: mm)

	$e = 0$ mm		$e = 200$ mm		$e = 1000$ mm	
	Sandy soil stratum	Soft soil	Sandy soil stratum	Soft soil	Sandy soil stratum	Soft soil
0°	−4.57	−12.65	−1.92	−3.61	0.12	0.24
45°	−1.38	−0.56	0.37	2.60	0.59	1.20
90°	5.41	12.78	2.15	3.45	0.11	−0.05
135°	1.72	0.31	−0.60	−2.73	−0.91	−1.53

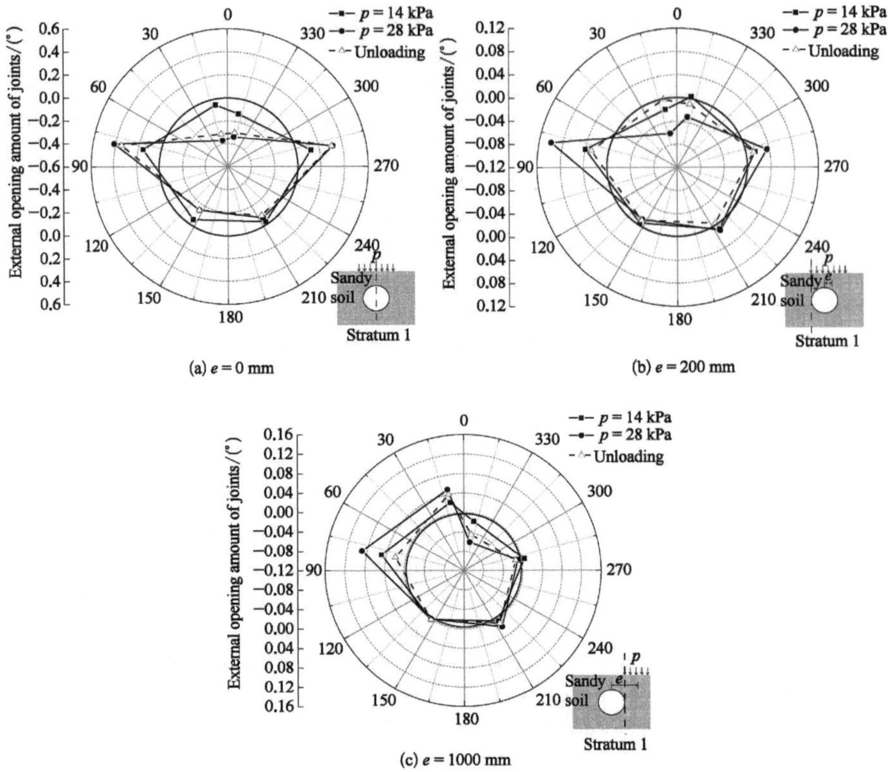

Fig. 3.34 Variation in tunnel horizontal convergence

deformation at lining segment joints. The interaction of these two types of deformation results in irrecoverable residual deformation of the lining segments after unloading. Therefore, engineering activities near tunnels should be strictly managed to prevent irreversible damage from sudden local loading.

(2) Joint opening

Using strain gauge displacement meters to measure the deformation of the inner and outer surfaces of tunnel joints enables the calculation of their outward opening angles (the outward opening amount). Figure 3.35 presents the joint opening of sandy soil tunnels under different loading eccentricities (positive values indicate outward opening; negative values indicate inward opening). At $e = 0$ mm, the opening angles are symmetrical: shoulder joints open outward, while others open inward. The outer opening angles of shoulder joints (78.75° and 281.25°) and inner opening angles of top joints (11.25° and 348.75°) are roughly equal. Upper joints exhibit greater deformation than lower ones, consistent with the numerical simulation results of Wang and Zhang (2013). As loading eccentricity increases, joint deformation decreases and becomes asymmetrical. At $e = 0$ mm, deformation primarily occurs in the left and right shoulder joints and top joints in a symmetrical pattern. Under

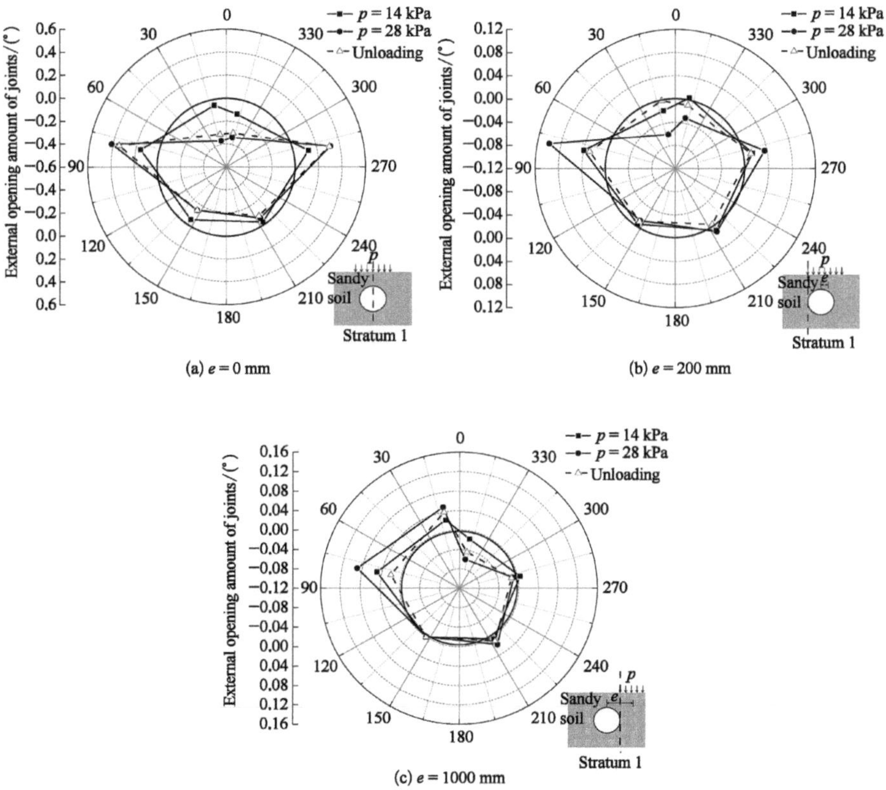

Fig. 3.35 Tunnel joints opening in sandy stratum

eccentric loading, it mainly happens in the shoulder and top joints opposite the load. Figure 3.36 illustrates the differences in deformation patterns between these two cases.

Figure 3.37 shows the deformation of tunnel joints in soft soil strata. Due to the large deformation, some strain gauges are damaged, so measurements are only taken for $e = 0$ mm and $e = 200$ mm. As shown in Fig. 3.38, the deformation of joints in soft soil strata is similar to that in sand strata, but with larger values. Also, there is still some residual deformation in the joints after unloading, indicating that the joint deformation in this test has entered the plastic development stage. Wang and Zhang (2013) numerically simulate the relationship between tunnel convergence deformation and joint opening under vertical loading. The results shows that the opening of the shoulder joints is about 10% of the tunnel convergence deformation. In this test, when $e = 0$ mm, the opening of the shoulder joints in the sand and soft soil strata tunnels is 4.4 and 1.9% of the maximum convergence deformation of the lining, respectively, which is different from the numerical simulation results in literature (Wang and Zhang 2013).

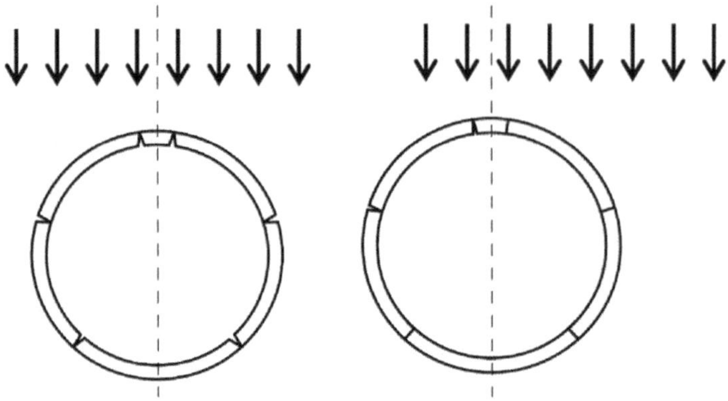

Fig. 3.36 Deformation of tunnel joints under loading from directly above and eccentric loading

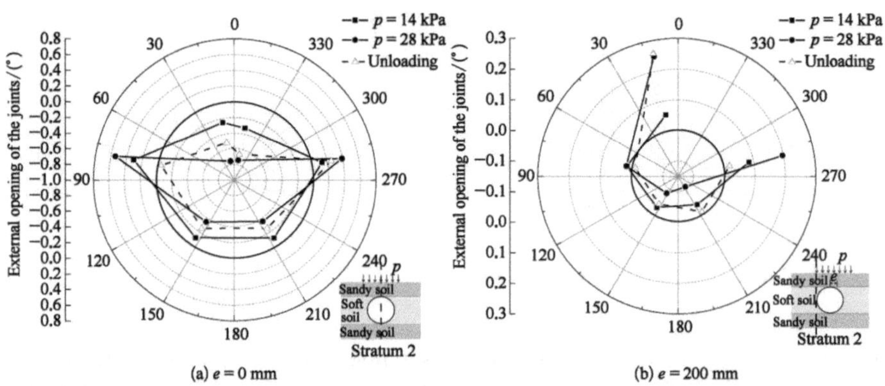

Fig. 3.37 External opening of tunnel joints in soft soils

(3) Earth pressure

Figure 3.38 shows the distribution of additional earth pressure on the tunnel's outer surface. When $e = 0$ mm, the sandy tunnel experiences relatively uniform additional stress distribution, while the soft soil tunnel's additional stress concentrates horizontally. Two factors contribute to this. Firstly, the greater convergent deformation in soft soil tunnels enhances the soil arch effect, making load redistribution at the tunnel top more pronounced. This leads to reduced vertical additional stress and increased horizontal stress. Secondly, the local additional stress from loading is about 27 kPa, and the overlying soil's self-weight stress is approximately 7 kPa. Comparing with the measured additional stress in the tunnel structure, it can be inferred that the tunnel's horizontal additional stress originates not only from ground local loading but also from the reaction force of the surrounding strata during horizontal deformation. As

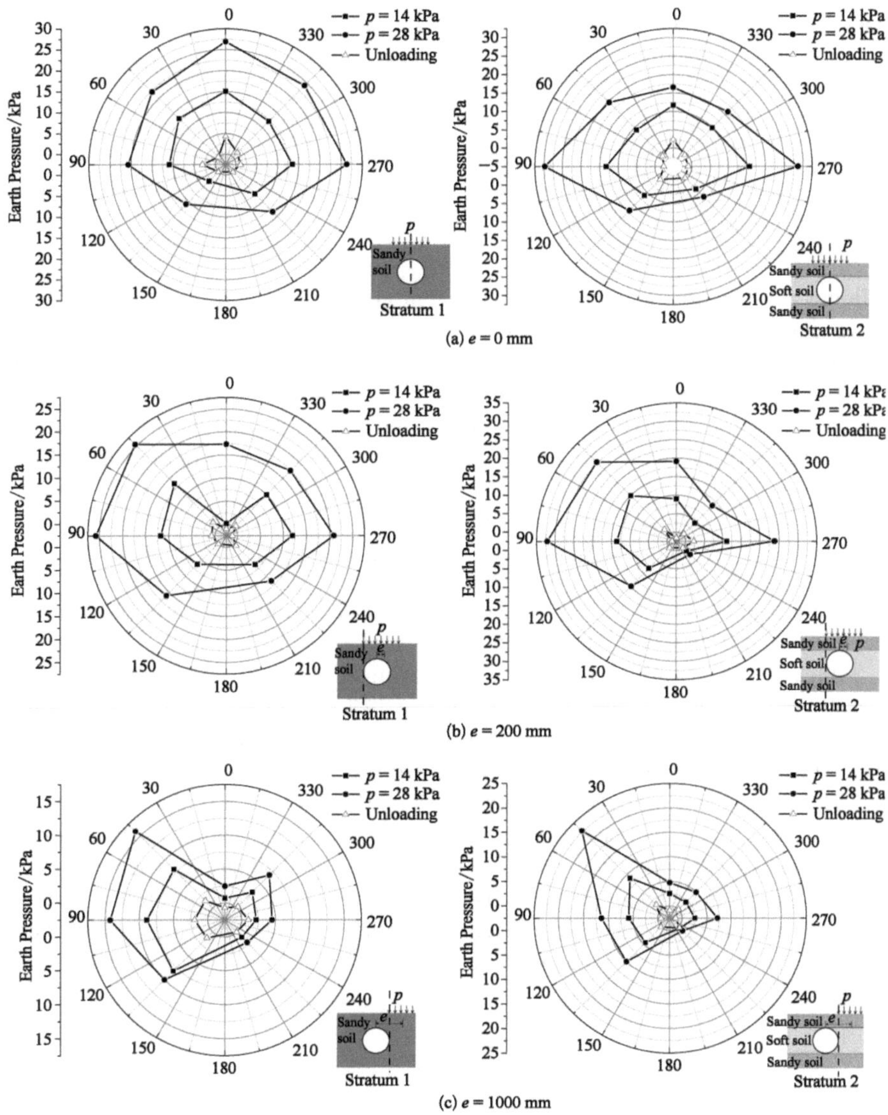

Fig. 3.38 Additional earth pressure on the outer surface of the tunnel structure

indicated in Fig. 3.34, the horizontal convergent deformation of the soft soil tunnel is roughly 2.4 times that of the sandy tunnel, resulting in greater horizontal additional stress in the former.

An increase in the loading eccentricity leads to a gradual decrease in the earth pressure on the tunnel surface as well as a change in the earth pressure distribution. In order to quantitatively characterize the concentration of earth pressure in the

Table 3.11 Earth pressure concentration factors for tunnel oblique cross-sections

	P/kPa	*e* = 0 mm	*e* = 200 mm	*e* = 1000 mm
Sandy soil	14	1.18	1.80	2.76
	28	1.05	1.54	2.91
Soft soil	14	1.19	2.11	3.09
	28	1.55	2.10	3.06

oblique cross-section, a coefficient of concentration of earth pressure in the oblique cross-section, *F*, is defined as follows.

An increase in the loading eccentricity causes a gradual reduction in the earth pressure on the tunnel surface and alters its distribution. To quantitatively describe the concentration of earth pressure on the oblique cross-section, a coefficient of concentration of earth pressure on the oblique cross-section, *F*, is defined as follows.

$$F = \frac{(S_{45} + S_{90})/2}{(S_0 + S_{135} + S_{225} + S_{270} + S_{315})/5}$$

where S_0, S_{45}, S_{90}, S_{135}, S_{225}, S_{270}, S_{315} represent the additional earth pressures on the sections at 0°, 45°, 90°, 135°, 225°, 270°, and 315°, respectively.

Table 3.11 presents how *F* varies with loading eccentricity and soil strata during loading. When *e* is constant, *F* values at $p = 14$ kPa and $p = 28$ kPa are similar, showing that local load magnitude mainly affects soil pressure magnitude, not its distribution. At $p = 28$ kPa, as *e* increases from 0 to 1000 mm, *F* rises from 1.05 to 2.91 in sandy soil tunnels and from 1.55 to 3.06 in soft soil tunnels. This means loading eccentricity increase leads to soil pressure concentration on the tunnel's diagonal section opposite the load. Comparing different soil strata, when $e = 0$ mm and 200 mm, *F* in soft soil tunnels is usually larger than in sandy soil tunnels. When $e = 1000$ mm, their *F* values are closer. This indicates that under small eccentricity loading, soil stratum properties significantly influence soil pressure distribution in tunnels.

3.4 Summary

In the finite element modeling section, ABAQUS software is used to analyze how loading size, tunnel depth, and stratum properties affect the deformation of tunnels under ground loading. Through parametric analysis, the following deformation patterns are identified:

① The transverse deformation of shield tunnels under ground loading mainly takes the forms of horizontal convergence, invert settlement and horizontal displacement. Horizontal convergence results from both the deformation of tunnel segments and their rotation around joints, whereas horizontal displacement mostly happens under eccentric loading.

② Small-scale local loading on the ground directly above soft-soil tunnels can induce considerable convergent deformation in shallow tunnels, which calls for special attention. Also, for the sake of tunnel structural safety and waterproofing, the ground loading should be controlled within 20 kPa when the tunnel is initially in good condition.

③ The region within 3 m of the tunnel contour is highly sensitive to engineering activities, which should be conducted with caution in this zone. Similarly, the area within the tunnel's main failure plane is sensitive, and large-scale engineering activities should be avoided there.

④ Horizontal convergence of the tunnel under ground loading is chiefly influenced by the soil properties of the stratum it traverses, while invert settlement is predominantly affected by the soil properties of the underlying stratum. For soft soil tunnels, settlement deformation should be prioritized when the underlying stratum is soft, whereas convergence deformation requires closer attention when the underlying stratum is hard.

The scaled-down model test section investigates the transverse deformation performance of subway tunnels in soft and hard soil strata under local ground loading. By controlling the loading eccentricity and the compression modulus of the soil strata traversed by the tunnel, comparative analyses of convergent deformation, joint opening, and additional soil pressure on the outer surface of the tunnels under different working conditions are performed. The main findings are summarized as follows:

① As the loading eccentricity increases, the tunnel's convergent deformation rapidly decreases before stabilizing, transitioning from transverse elliptical deformation to oblique elliptical deformation. The convergence in soft soil tunnels is greater than in sand tunnels, and increasing the loading eccentricity reduces convergence more significantly in soft soil tunnels.

② Increasing the loading eccentricity reduces the tunnel joint deformation. When loaded from directly above, deformation is mainly symmetric at the tunnel's left and right shoulders and top joints. Under eccentric loading, deformation primarily occurs at the shoulders and top joints on the side opposite to the load. Soft soil tunnels experience larger joint deformations than sandy soil tunnels.

③ Under loading, additional earth pressure distribution in hard soil tunnels is relatively uniform, while in soft soil tunnels, it is predominantly concentrated in the horizontal direction. With the increase of loading eccentricity, the additional earth pressure gradually decreases, and the location of maximum pressure shifts to the oblique section on the side opposite to the load. Sudden localized ground loading can cause significant damage to tunnels, which is difficult to recover after unloading.

In conclusion, to control the shield tunnel's transverse deformation and maintain its good service condition, loading activities on the surface above soft-soil tunnels should strictly follow the principle of "prevention first, reducing and distancing loads, advance protection, and real-time monitoring".

References

Fang YQ (2022) Mechanism of transverse deformation and protection/restoration technologies for metro shield tunnels under surcharge loading. PhD dissertation, Tongji University, Shanghai

Gong C, Ding W, Soga K et al (2019) Failure mechanism of joint waterproofing in precast segmental tunnel linings. Tunn Undergr Space Technol 84(02):334–352

Huang H, Shao H, Zhang D et al (2017) Deformational responses of operated shield tunnel to extreme surcharge: a case study. Struct Infrastruct Eng 13(03):345–360

Huang DW, Zhou SH, Feng QS et al (2018) Design of model test for interaction between shield tunnel and surrounding strata. J China Railw Soc 40(06):127–135

Huang DW, Zhou SH, Feng QS et al (2019) Experimental study on influence of surface surcharge on existing shield tunnels in soft and hard strata. Chin J Geotechn Eng 41(05):942–949

Liu JH, Hou XY (1991) Shield tunneling method. China Railway Publishing House, Beijing

Wang RL, Zhang DM (2013) Mechanism and control index of transverse deformation of shield tunnels in soft soil under overload. Chin J Geotechn Eng 35(06):1092–1101

Yi H, Qi T, Qian W et al (2019) Influence of long-term dynamic load induced by high-speed trains on the accumulative deformation of shallow buried tunnel linings. Tunn Undergr Space Technol 84(02):166–176

Yuan Y, Bao Z, Yu HT et al (2017) Multi-point shaking table test of shield tunnel considering traveling wave effect. China J Highw Transp 30(08):174–182

Yuan Y, Yang Y, Zhang S et al (2020) A benchmark 1 g shaking table test of shallow segmental mini-tunnel in sand. Bull Earthq Eng 18(11):5383–5412

Chapter 4
Numerical Analysis of Shield Tunnel Deformation Under Upper Unloading

4.1 Introduction

With the development of urban underground spaces and river dredging, there are increasingly more cases of foundation pit unloading above shield tunnels. The primary impacts on the tunnel are significant uplift and minor convergence deformations. Extensive engineering practice indicates that the impact of foundation pit construction on underlying tunnels is a highly complicated process, with factors such as foundation reinforcement, foundation pit support systems, excavation size, and foundation pit exposure time all playing crucial roles in the tunnel heave deformation. Although measured data from some engineering cases show relatively small transverse convergence deformation, often manifesting as vertical ovalization deformation, its impact should not be overlooked. While each project undergoes detailed design evaluation before construction, uncertainties in assessing the impact on the strata and construction have led to significant uplift and convergence deformations in some cases. This highlights that research on such conditions still lags far behind engineering practice and practical needs.

Scholars conduct comparative studies primarily through theories, field experiments, and numerical simulations. Research mainly focuses on the mechanisms and patterns of bottom heave deformation caused by excavation, the influence rules of reinforcement in adjacent tunnel soil and pit bottom, the impact of temporal and spatial effects during excavation on tunnel deformation, and the summary of empirical formulas for deformation-pattern curves based on engineering cases. These studies have yielded certain results and predictive methods to guide engineering design and construction.

Based on existing research findings and recent understanding of soil constitutive models in Shanghai's soft soil areas, this chapter conducts a specialized study on the impact of tunnel deformation due to unloading from excavation directly above

R. Wang et al., *Impact Analysis and Control Strategies of Overlying Loading/Unloading Effects on Soft Soil Shield Tunnels*, Key Technologies for Tunnel Construction under Complex Geological and Environmental Conditions,
https://doi.org/10.1007/978-981-95-1462-5_4

subway tunnels. Factors such as excavation size, soil reinforcement, engineering protection piles, extreme working conditions, and time restrictions on excavation exposure are introduced. Through finite element numerical simulation analysis, the spatial and temporal effects of excavation on tunnel deformation are discussed. By integrating tunnel monitoring, the chapter explores the pattern of the impact of excavation directly above tunnels on tunnel deformation, aiming to provide references for future engineering practices.

4.2 Determination of Soil Constitutive Models and Computational Parameters

The determination of appropriate soil constitutive models is crucial for geotechnical finite element analysis. This section provides a brief introduction to the characteristics of soil constitutive models and discusses their applicability in sensitive geotechnical engineering environments in the Shanghai area, laying a theoretical foundation for the subsequent research studies in this chapter.

4.2.1 Soil Constitutive Models

Up to now, hundreds of soil constitutive models have been proposed, but each model reflects one or several types of soil. Due to the complexity of soil deformation behavior, each model has its own range of application and limitations. Currently, the main models for soil include: elastic models such as the linear elastic model and the Duncan-Chang model; elastic-perfectly plastic models such as the Mohr–Coulomb (MC) model and the Drucker-Prager (DP) model; hardening elastoplastic constitutive models, such as the Modified Cam-Clay (MCC) model, the Hardening Soil (HS) model, and the small strain model.

(1) Elastic constitutive models

(1) The linear elastic model

The linear elastic model is one of the most fundamental and simple mechanical models. The constitutive relationship of linear elastic materials follows the generalized Hooke's law, meaning the stress–strain exhibits a linear relationship during loading–unloading, with no residual strain after unloading. The stress–strain relationship is described by two parameters: Elastic modulus of elasticity E and Poisson's ratio ν. While the linear elastic model represents the simplest stress–strain relationship, it cannot depict many important characteristics of soil, such as plastic deformation, dilatancy, etc.

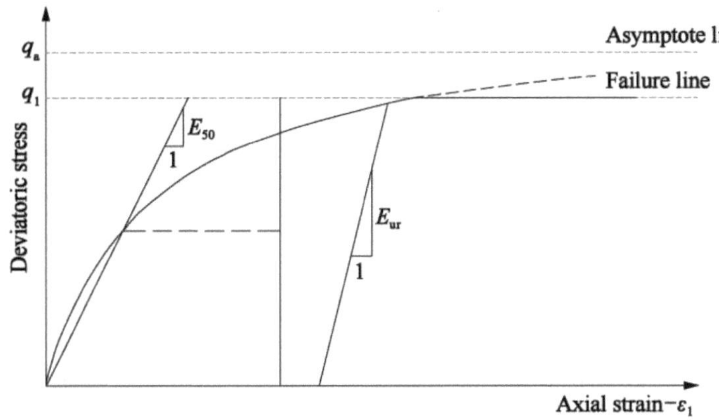

Fig. 4.1 Typical stress–strain relationship in triaxial tests

(2) The non-linear Duncan-Chang model

This model is a type of nonlinear elastic model based on the incremental form of the generalized Hooke's law. It is capable of reflecting the nonlinearity in the stress–strain relationship of soils. There are a total of eight model parameters with clear physical meanings, which can be determined through conventional triaxial tests. Its ease of use in numerical calculations has led to widespread application.

In 1963, Kondner (1963), based on a large number of triaxial test stress–strain curves for soil, proposed that a hyperbola could fit the general soils triaxial test curves, as shown in Fig. 4.1. In this figure, q_a represents the limit value of deviatoric stress, and q_f the failure value of deviatoric stress. The formula:

$$\sigma 1 - \sigma 3 = \frac{\varepsilon_a}{a + b\varepsilon_a} \tag{4.1}$$

In the equation: a and b are experimental constants; for conventional triaxial compression tests, the axial strain $\varepsilon_a = \varepsilon_1$.

In 1970, Duncan and Chang (1970), based on the hyperbolic stress–strain relationship, put forward an incremental elastic model, also known as the Duncan-Chang model.

The Duncan-Chang model, with its relatively simple mathematical equation, captures the non-linear elasticity of geomaterials. However, the model remains rooted in elastic theory without incorporating plasticity concepts. It cannot account for the effects of different stress paths, nor does it contemplate the factors of the compressibility, dilatancy and strain-softening of geomaterials, as well as the impacts of the intermediate principal stress on strength and deformation.

(2) Elastic-perfectly plastic constitutive models

(1) The Mohr–Coulomb (MC) model

The Mohr–Coulomb (MC) model is an elastic-perfectly plastic model that integrates Hooke's law with the Coulomb's failure criterion. Soil is assumed to behave as a linear elastic material until it reaches its failure strength, after which it becomes perfectly plastic. The formula for the failure criterion is:

$$\tau_{\mathrm{f}} = \sigma \tan\varphi + c \tag{4.2}$$

The MC model has five model parameters, namely two parameters that control elastic behavior (Elastic modulus E and Poisson's ratio v), and three parameters that control plastic behavior (effective cohesion c, effective angle of internal friction φ, and dilation angle ψ).

The MC model is relatively simple, which offers clearly defined parameters that effectively characterize soil failure behavior. However, it fails to accurately represent the actual deformation mechanisms prior to failure, since it assumes that the stress–strain relationship of soil follows Hooke's Law before reaching the shear strength. The model uses the same loading and unloading modulus and does not account for the effects of stress paths, potentially leading to significant discrepancies in wall horizontal displacement, pit bottom rebound, and external soil settlement during excavation unloading simulation, which largely deviates from the actual situation. Although widely used, the MC model's hexagonal yield surface in the deviatoric plane can cause convergence difficulties in plastic analysis due to the non-unique flow direction at the corners.

(2) The Drucker-Prager (DP) model

Drucker and Prager (1952) combined the Coulomb yield criterion which ignores the effect of the intermediate principal stress σ_2, with the von Mises yield criterion which ignores the effect of hydrostatic pressure P, thereby introducing an extended Mises perfectly plastic model, known as the DP model (Fan et al. 2014). In 1957, Drucker et al. (1957) further developed the concept of work-hardening in soil plasticity mechanics, suggesting that hydrostatic pressure affects soil yield. They adapted the original DP model by adding a hemispherical "cap" yield surface to its open end. This concept marked the earliest proposition of the cap model, laying the groundwork for the application of work-hardening theory in soil mechanics.

The yield criterion in the DP model adopts the generalized Mises yield criterion, whose formula is:

$$F = \sqrt{J_2} - \alpha I_1 - k = 0 \tag{4.3}$$

In the equation: J_2 is the second invariant of the deviatoric stress; I_1 is the first invariant of the stress tensor; the parameters α and k can be determined from the yield stress data characterizing the material's cohesion c and friction angle φ. For triaxial compression, the parameters α and k can be expressed as:

$$\alpha = \frac{2\sin\varphi}{\sqrt{3}(3 - \sin\varphi)} \tag{4.4}$$

$$k = \frac{6c\cos\varphi}{\sqrt{3}(3 - \sin\varphi)} \tag{4.5}$$

Under plane strain conditions, the relationship between the parameters α and k and the cohesion c and friction angle φ is as follows:

$$\alpha = \frac{\sin\varphi}{\sqrt{3}} \tag{4.6}$$

$$k = \frac{2c\cos\varphi}{\sqrt{3}} \tag{4.7}$$

The yield surface represented by the DP model's yield criterion forms a conical surface in the principal stress space. The DP model modifies to the yield surface function of the MC model. It replaces the hexagonal pyramid yield surface of the MC model with a conical yield surface, which simplifies programming and numerical calculations. However, the shortcomings of the MC model persist in the DP model as well.

(3) Hardening elasto-plastic constitutive models

(1) The Modified Cam-Clay Model

The modified cam-clay model was developed by Roscoe et al. (1963), based on the drained and undrained triaxial tests conducted on normally consolidated and slightly over-consolidated clay samples.

The Cambridge (Cam-Clay) model was proposed based on the Critical State Theory. Subsequently, Roscoe and Burland further developed the Modified Cambridge (MCC) model (Roscoe and Burland 1968). The MCC model effectively elucidates the elastoplastic deformation characteristics of soils, both theoretically and experimentally, making it one of the most widely applied constitutive models for soft soil.

The Cam-Clay model assumes that soft soil undergoes isotropic compression under drained conditions, the specific volume v and ln$p$' form a straight line, known as the virgin compression line, as illustrated in Fig. 4.2. Upon unloading, the sample swells and the specific volume v and ln$p$' form a swelling line. The effective mean stress p' is defined as:

$$p' = \frac{1}{3}(\sigma_1' + \sigma_2' + \sigma_3') \tag{4.8}$$

The specific volume v is defined as:

$$\nu = 1 + e \tag{4.9}$$

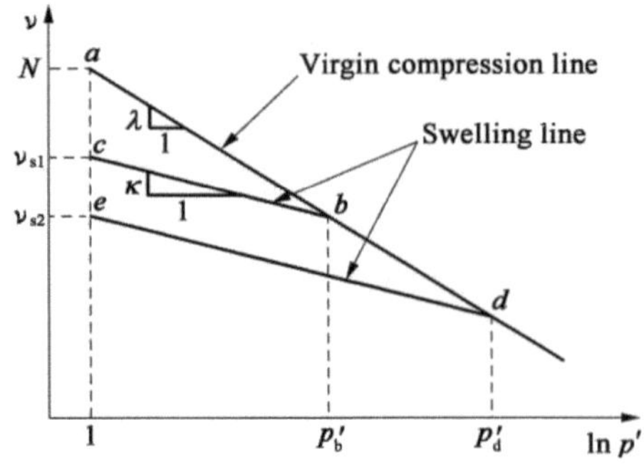

Fig. 4.2 Virgin compression line and swelling line

The virgin compression line is defined by the equation:

$$v = N + \lambda \ln p' \tag{4.10}$$

The swelling line is defined by the equation:

$$v = v_s + \kappa \ln p' \tag{4.11}$$

In the equation: λ is the slope of the virgin compression line in the v-lnp' plane, while κ is the slope of the swelling line. N represents the specific volume of the virgin compression line at unit pressure, while v_s represents the specific volume corresponding to unloading to unit pressure on the swelling line.

Meanwhile, under continuous shear stress, the volumetric deformation and stress remain constant, but the shear deformation continues to increase until failure. The trajectory of the failure point in the shear test defines the Critical State Line (CSL). CSL is parallel to the virgin compression line in v-lnp' plane, as shown in Fig. 4.3. The relationship between the parameter N and Γ can be expressed by:

$$\Gamma = N - (\lambda - \kappa)\ln 2 \tag{4.12}$$

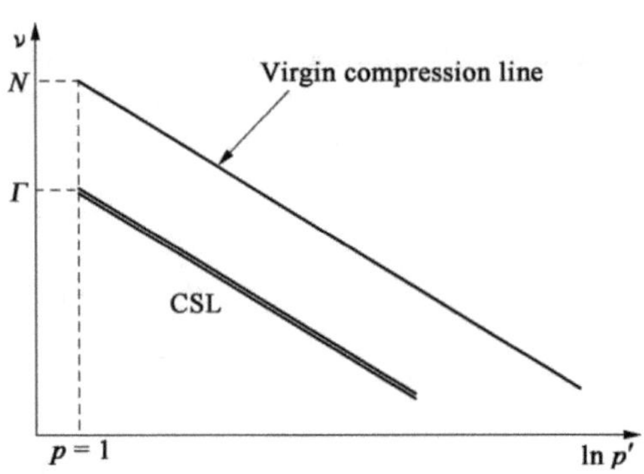

Fig. 4.3 The critical state line in v-lnp' plane

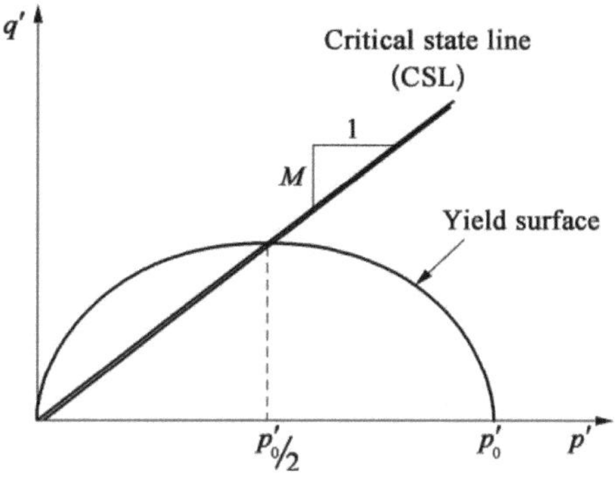

Fig. 4.4 The yield surface of MCC model in $p' - q'$ plane

In the equation: The parameter Γ is the specific volume of the CSL at unit pressure.

The yield surface of MCC model in p'-q' plane is illustrated in Fig. 4.4.

The model features well-defined fundamental concepts, considering the yield characteristics of geomaterials under hydrostatic pressure, as well as their contractive and dilative behaviors. It is well-suited for normally consolidated clays and lightly over-consolidated clays. The computational parameters can be easily obtained through experiments, facilitating its promotion in geotechnical engineering. However, the MCC model has its limitations. Its yield surface represents only the isochoric surfaces of plastic volumetric strains, employing plastic volumetric strain as the hardening parameter, thus inadequately addressing shear deformation. It can only reflect soil contractancy, but not dilatancy. In addition, the model does not consider the variation characteristics of soil stiffness within a small strain range.

(2) The Hardening Soil (HS) model

The HS model, proposed by Schanz et al. (2000) based on Vermeer's double-hardening model, is an isotropic hardening elastoplastic model. Similar to the Duncan-Chang model, the HS model assumes a hyperbolic relationship between shear stress q and axial strain ε_1 in triaxial drained shear tests. However, the HS model replaces the elastic theory of the Duncan-Chang model with plasticity theory and considers dilatancy of soil. Moreover, the model incorporates soil dilatancy and introduces a yield cap, enabling simultaneous consideration of both shear hardening and compression hardening of soils. Additionally, the HS model incorporates an unloading–reloading modulus E_{ur}, accounting for the unloading effects during excavation.

The model consists of 11 parameters which are defined in Table 4.1.

The yield curve in primary loading for a standard consolidated drained triaxial test can be described as:

$$-\varepsilon 1 = \frac{1}{2E_{50}} \frac{q}{1 - q/q_{\mathrm{a}}} q < q_{\mathrm{f}} \tag{4.13}$$

Table 4.1 Definitions of the HS model parameters

	Definition
c'	Effective cohesion of soil
φ'	Effective internal friction angle of soil
K_0	Coefficient of lateral earth pressure "at rest" of normally-consolidated soil
ψ	Dilatancy angle of soil
m	Exponential power of stiffness stress level
p^{ref}	Reference stress
E_{50}^{ref}	Reference secant modules from triaxial drained shear tests
$E_{\text{ur}}^{\text{ref}}$	Reference loading–unloading modules from triaxial drained shear tests
$E_{\text{oed}}^{\text{ref}}$	Reference tangent modulus for oedometer tests
R_{f}	Failure ratio
ν_{ur}	Loading–unloading Poisson's ratio

In the equation: The q_{a} is the ultimate value of shear strength, while the q_{f} is deviatoric stress at failure. The parameter E_{50} is secant modulus corresponding to 50% of ultimate load when primary loading, which is given by the equation.

$$E_{50} = E_{50}^{\text{ref}}\left(\frac{c' \cdot \cos\varphi' - \sigma_3' \sin\varphi'}{c' \cdot \cos\varphi' + p^{ref} \sin\varphi'}\right)^m \tag{4.14}$$

In the equation: E_{50}^{ref} is the reference secant modulus corresponding to the reference stress p^{ref}.

The reference modulus for loading–unloading $E_{\text{ur}}^{\text{ref}}$ and tangent modulus $E_{\text{oed}}^{\text{ref}}$ can be expressed as

$$E_{\text{ur}} = E_{\text{ur}}^{\text{ref}}\left(\frac{c' \cos\varphi' - \sigma_3' \sin\varphi'}{c' \cos\varphi' + p^{ref} \sin\varphi'}\right)^m \tag{4.15}$$

$$E_{\text{oed}} = E_{\text{oed}}^{\text{ref}}\left(\frac{c' \cos\varphi' - \sigma_3' \sin\varphi'}{c' \cos\varphi' + p^{ref} \sin\varphi'}\right)^m \tag{4.16}$$

The HS model adheres to the Mohr–Coulomb failure criterion, and the failure value of the soil's deviatoric stress, q_{f}, can be expressed as:

$$q_{\text{f}} = (\text{ctan}\varphi - \sigma_3')\frac{2\text{sin}\varphi}{1 - \text{sin}\varphi} q_{\text{a}} = q_{\text{f}}/R_{\text{f}} \tag{4.17}$$

The HS model is an isotropic hardening elasto-plastic model that covers both shear hardening and compression hardening of soils. Additionally, the model introduces the loading–unloading modulus E_{ur}, considering the effects of unloading during excavation, making it suitable for analyzing deformation impacts on surrounding environments in foundation pit engineering. Nevertheless, it fails to consider the variation in soil stiffness within the small strain range, thus limiting its application to sensitive environments for deep foundation pit engineering.

(3) The Hardening Soil Model with small strain stiffness

Benz (2006) enhanced the HS model by incorporating a modified version of the Hardin-Drnevich shear modulus equation (Hardin and Drnevich 1972), thereby accounting for the non-linear variation of soil stiffness with strain in the small strain region (Benz 2006). Experiments have shown that the stress–strain curve under the small strain can be simulated by a hyperbolic. The shear modulus equation given by the Hardin-Drnevich model is

$$\frac{G_s}{G_0} = \frac{1}{1 + \left|\frac{\gamma}{\gamma_r}\right|} \tag{4.18}$$

The ultimate shear strain γ_r is defined as:

$$\gamma_r = \frac{\tau_{max}}{G_0} \tag{4.19}$$

In the equation: where τ_{max} is the shear stress at failure.

To avoid the incorrect use of a larger ultimate shear strains, it is recommended to use the shear strain $\gamma_{0.7}$, which corresponds to a reduction of the secant modulus G_s to 70% of its initial value, as a substitute for γ_r. Equation (4.19) can be modified as follows:

$$\frac{G_s}{G_0} = \frac{1}{1 + \alpha\left|\frac{\gamma}{\gamma_{0.7}}\right|} a = 0.385 \tag{4.20}$$

Benz assumed that G_0 and the initial shear modulus at the reference stress p^{ref} have the following relationship:

$$G_0 = G_0^{ref}\left(\frac{c\cot\varphi + \sigma_3}{c\cot\varphi + p^{ref}}\right)^m \tag{4.21}$$

Similarly, E_0 and the initial elastic modulus at the reference stress p^{ref} have the following relationship:

$$E_0 = E_0^{ref}\left(\frac{c\prime\cot\varphi\prime + \sigma\prime_3}{c\prime\cot\varphi\prime + p^{ref}}\right)^m \tag{4.22}$$

The HSS model accounts for the hardening characteristics of soft clay, distinguishes between loading and unloading, and incorporates stress history and stress path-dependent stiffness. It also considers the nonlinear variation of soil stiffness with strain in the small strain range.

The experimental determination method for the soil initial modulus G_0 is commonly measured using dynamic methods, including in-situ and laboratory dynamic tests. In-situ dynamic tests contain Cross Hole Seismic Test (CHST) and Seismic Cone and Seismic Flat Dilatometer (SCPT), etc. Laboratory dynamic tests include resonant column and bender elements methods. The resonant column test is one of the most reliable methods for determining the maximum shear modulus of soil. Its principle involves applying axial or torsional harmonic vibrations to one end of the soil sample, generating stress waves that travel to the other end and are reflected back. By adjusting the excitation frequency to achieve maximum strain amplitude at the vibrating end of the sample, resonance occurs in the sample system. The measured resonance frequency at this point is the natural frequency f_{n} of the soil sample system, and then the dynamic shear modulus G_d can be calculated using Eq. (4.23).

$$G_{\mathrm{d}} = \rho\left(\frac{2\pi f_{\mathrm{nT}} H}{\beta_{\mathrm{T}}}\right)^2 \tag{4.23}$$

In the equation: G_{d} represents the dynamic shear modulus of the sample (MPa); ρ denotes the mass density of the sample (kg/m^3); f_{nT} is the torsional vibration resonance frequency of the system (Hz); H is the height of the sample after consolidation (m); β_{T} is the eigenvalue of the torsional vibration frequency equation.

Bender element method is another effective method for measuring the maximum shear modulus of soils. Bender elements are piezoelectric electromechanical sensors that can convert electrical energy into mechanical motion. By installing bender elements at the base and cap of a triaxial pressure chamber and inputting an electric current, the elements bend and generate shear waves that pass through the soil sample, with the speed of these shear waves denoted as V_{s}. The transmission of the waves causes the receiving element to vibrate, and the oscilloscope captures the electrical signal. The shear wave speed, V_{s}, is determined by the time taken for the shear wave to travel between the source and receiving elements. The very small strain shear modulus $G_{\max}$ can be calculated using Eq. (4.24):

$$G_{\max} = \rho V_{\mathrm{s}} \tag{4.24}$$

In the equation: P is soil sample's density.

The HSS model, in addition to all the parameters of the HS model (as listed in Table 4.1), also includes two small-strain parameters G_0^{ref} and $\gamma_{0.7}$, totaling 13 parameters. Among these, the reference initial shear modulus G_0^{ref} can be replaced by the reference initial elastic modulus E_0^{ref}, $E_0^{\mathrm{ref}} = 2(1+v)G_0^{\mathrm{ref}}$, where v is the soil's Poisson's ratio.

(4) Determination of soil constitutive models

The introduction on soil constitutive models reveals that each model captures only certain aspects of soil behavior, making it challenging to develop a universal mathematical model that fully describes all soil characteristics. When selecting a foundation model, it is essential to make targeted choices based on the type of soil and the characteristics of engineering projects. Comparatively, the HSS model not only considers the compressibility and dilatancy of soft clay, distinguishing between loading and unloading stiffness, but also recognizes that its stiffness depends on stress history and stress path. Additionally, it accounts for the characteristic of shear modulus decay with increasing strain in the small strain range, making it suitable for analyzing geotechnical engineering in Shanghai's complex and sensitive environments. The following sections will focus on the parameter determination methods and applications of the HSS model for Shanghai soft soils, as detailed in *Shanghai Housing and Urban-Rural Construction Management Committee* (2018).

4.2.2 *Experimental Study on Parameters of HSS Model for Shanghai Soft Soils*

Due to numerous parameters of the HSS model, it is difficult to obtain a complete set of HSS model parameters. The major parameters of the HSS model for typical soil layers in Shanghai, including clay of layer ②, muddy silty clay of layer ③, muddy clay of layer ④ and silty clay of layer ⑤, were obtained through laboratory tests in this section. The applicability of the HSS model and the proposed parameter determination methods in Shanghai are verified through engineering case studies (Liang et al. 2017).

(1) Experimental process

(1) Experimental purposes and instruments

The parameters of the HSS model such as K_0, ψ, p^{ref}, ν_{ur}, and m can be evaluated with reference to existing research findings. The specific evaluation methods and sources of literature are listed in Table 4.2. This section focuses on determining the other 8 parameters. For Shanghai's typical soil layers: clay of layer ②, muddy silty clay of layer ③, muddy clay of layer ④ and silty clay of layer ⑤. These parameters are determined through four sets of experiments, which are as follows: conventional triaxial consolidated drained shear tests to obtain c', φ', R_{f} and E_{50}^{ref}; triaxial consolidated drained loading and unloading tests to obtain $E_{\text{ur}}^{\text{ref}}$; standard consolidated tests to obtain $E_{\text{oed}}^{\text{ref}}$; resonant column tests to obtain small strain parameters G_0^{ref} and $\gamma_{0.7}$.

The triaxial consolidated drained shear tests and the triaxial consolidated drained loading–unloading tests employ the Unsaturated Triaxial Testing of Soil (UNSAT) manufactured by GDS Instruments, UK. The resonant column tests are performed using the V. P. Drnevich resonant column apparatus. The standard oedometer tests

Table 4.2 Evaluation of partial parameters of HSS model

	Evaluation
K_0	$(1 - \sin\phi')$
ψ	For sandy soil, if φ' is greater than 30°, take (φ' − 30°); if φ' is less than 30°, then it is taken as 0°. For cohesive soil, it is generally taken as 0°
p^{ref}	100 kPa
ν_{ur}	0.2
m	For sandy and silty soil, a value of 0.5 is typically used; for cohesive soil, the value ranges from 0.5 to 1.0

are conducted using conventional oedometer apparatus, with the test procedures following the standards for geotechnical testing methods.

(2) On-site sampling

The soil samples are taken from a deep foundation pit construction site in Pudong, Shanghai, at depths of 3, 7, 15, and 20 m, corresponding to the four typical soil layers (②③④⑤) in Shanghai. The basic physical and mechanical parameters of the soil samples from each layer are shown in Table 4.3. All samples required for the aforementioned four sets of tests are cylindrical in shape, but vary in size as follows: for the laboratory triaxial tests (diameter 39.1 mm, height 80 mm), standard consolidation tests (diameter 61.8 mm, height 20 mm), and resonant column tests (diameter 35.7 mm, height 80 mm). First, the soil samples are prepared to the required sizes, and then their actual sizes are measured using calipers.

(2) Experimental procedures

(1) The conventional triaxial consolidated drained shear test

① Back pressure saturation: The "Saturation Ramp" test module is used, where back pressure saturation is applied to saturate the specimen. A back pressure of 100 kPa is applied at the top of the specimen while a confining pressure of 110 kPa is applied around the specimen, causing air bubbles within the soil to shrink. This process is maintained for approximately 3 h.

Table 4.3 Physical parameters of soil layers

	Water content w/%	Unit weight γ/(kN/m^3)	Initial void ratio e_0	Sampling depth/m
② Clay	28.34	18.75	0.87	3
③ Muddy silty clay	29.00	18.14	0.93	7
④ Muddy clay	32.53	17.29	1.11	15
⑤ Silty clay	24.34	19.29	0.75	20

② B-value check: The "B-Check" module is selected, keeping the back pressure constant and increasing the confining pressure by 30 kPa. The pore water pressure coefficient B is measured under undrained conditions. If $B = \Delta u/\Delta\sigma > 95\%$, the specimen is considered saturated.

③ Consolidation: The "Consolidation" test module is used to isotropically consolidate the specimen by setting the confining pressure σ_3. The consolidation stabilization time is set to 36 h. In this test, three different confining pressures σ_3 are applied for each soil layer: 100, 200, and 300 kPa for layers ③-⑤. Layer ②, being shallower, is subjected to confining pressures of 70, 100, and 200 kPa.

④ Drained shear: The "Advance loading" test module is selected, maintaining the confining pressure σ_3 unchanged. Drained shear is conducted under strain rate control at a rate of 0.0056 mm/min to ensure sufficient time for pore water drainage. The test is halted when the specimen's strain value reaches 20%.

(2) The triaxial consolidated drained loading–unloading test

Only Step ④ differs from the triaxial consolidated-drained shear test. The specific procedure is as follows: The specimen undergoes axial loading–unloading-reloading. The "Advance Loading" test module is used, setting the confining pressure σ_3 to the reference confining pressure p^{ref} (100 kPa). Based on the triaxial test results of the same soil layer under the reference confining pressure, the anticipated failure deviatoric stress of the specimen is estimated. Firstly, axial loading is applied up to 40% of the estimated failure deviatoric stress. Upon reaching the target load, axial unloading is performed down to zero. And then axial loading is reapplied up to 60% of the estimated failure deviatoric stress.

(3) Resonant column test

① Consolidation: After the specimen is mounted on the resonant column device, a consolidation pressure σ_3 is applied and the drainage valve is opened until the specimen is fully consolidated. The consolidation pressure σ_3 corresponds to the vertical effective overburden stress at the sampling depth of each soil layer, which are 30, 70, 150, and 200 kPa for layers ②–⑤ respectively.

② Vibration: After applying a vibratory force to the specimen, the vibratory frequency is continuously changed from a low value, gradually increasing until the system resonates. The shear strain amplitude and resonance frequency at the time of resonance are recorded to calculate the specimen's dynamic shear modulus G_{d}. After each level of vibratory force testing, the vibratory force is incrementally increased. The test is continued to measure the modulus G_{d} at higher strain amplitudes.

③ At the end of the test, the pressure is released, the outer cover of the pressure chamber is removed, and the specimen is dismantled.

(3) Experimental results

(1) The results of the standard triaxial consolidated drained shear test

The standard triaxial consolidated drained shear test provides the reference secant modulus E_{50}^{ref}, and failure ratio R_{f}, as well as its strength parameters, including the effective cohesion (c') and the effective angle of internal friction (φ').

① E_{50}^{ref} and R_{f}

Under the confining pressure σ_3 of 100 kPa, the relationship between deviatoric stress $q(q = \sigma_1 - \sigma_3)$ and axial strain ε_1 of the soil samples from layers ②–⑤ is shown in Fig. 4.5.

As can be seen from Fig. 4.5, under a confining pressure σ_3 of 100 kPa, the relationship between deviatoric stress and axial strain for soil samples from layers ②–⑤ is quite similar. Initially, the deviatoric stress increases with the increase of axial strain when the axial strain is small. However, once the axial strain exceeds a certain range, the deviatoric stress no longer increases but remains constant or slightly decreases. The failure value q_{f} is determined by taking either the peak stress value from the curve or the deviatoric stress corresponding to an axial strain of 15%. The q_{f} values for layers ②–⑤ are respectively 223.7, 307.6, 185, and 317.3 kPa. The reference modulus E_{50}^{ref} corresponds to the secant modulus at 50% of the ultimate load q_{f}. The slope of the line connecting the origin and the point corresponding to $0.5q_{\mathrm{f}}$ represents the reference

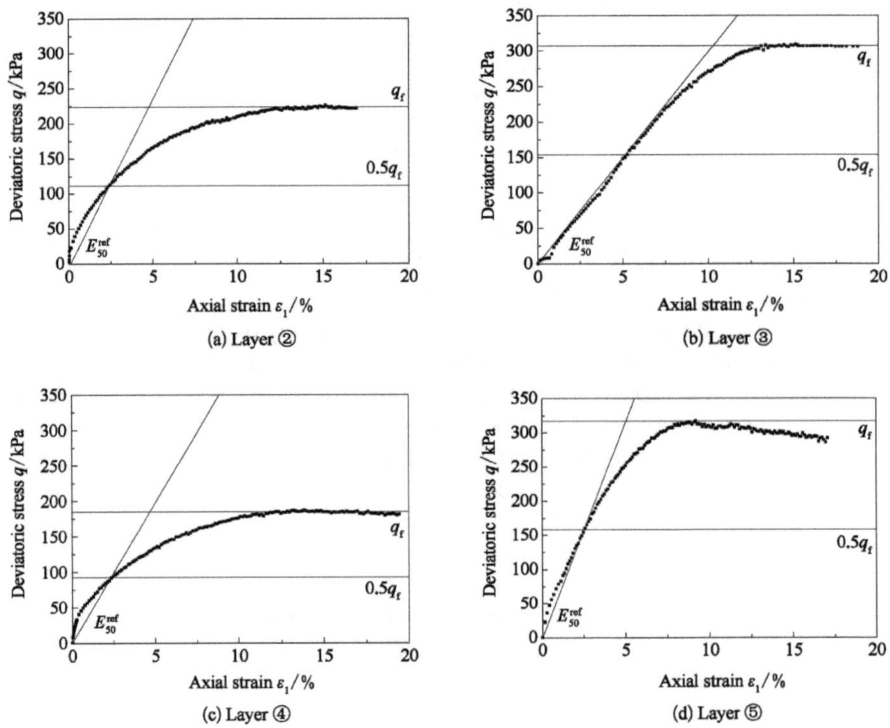

Fig. 4.5 Strain–stress curves of triaxial CD tests on soil layers (②–⑤)

modulus E_{50}^{ref}. Therefore, at a reference stress p^{ref} of 100 kPa, the reference modulus E_{50}^{ref} for layers ②–⑤ are 4.7, 3.0, 3.9, and 6.4 MPa, respectively.

The fundamental concept of the Hardening-Soil model is that the relationship between vertical strain ε_1 and deviatoric stress q under triaxial loading follows a hyperbolic relation.

$$-\varepsilon_1 = \frac{q}{2E_{50}(1 - q/q_a)} q < q_f \tag{4.25}$$

By changing the equation, the hyperbolic stress–strain relationship q–ε_1 can be transformed into linear relationship ε_1/q–ε_1:

$$\frac{\varepsilon_1}{q} = \frac{\varepsilon_1}{q_a} - \frac{1}{2E_{50}} \quad q < q_f \tag{4.26}$$

The actual stress–strain relationship does not fully conform to the assumed hyperbolic relationship. Especially in the part of the beginning phase and the phase nearing failure, the experimental data sometimes deviate from the linear relationship when converting the hyperbolic q–ε_1 stress–strain relationship into the linear ε_1/q-ε_1 relationship. To minimize human factors and achieve better overall conformity, the line is drawn through points at stress levels $s = 70\%$ and $s = 95\%$, where $s = q/q_f$. The ε_1/q-ε_1 curves (Fig. 4.6) are then plotted, and the values of q_a and R_f are obtained from the slope of this line (Table 4.4).

② Values of c' and φ'

To obtain the values of the soil's c' and φ', this experiment takes three different confining pressures σ_3 for each soil layer. The Mohr's circles corresponding to these confining pressures are depicted in Fig. 4.7. By drawing the common tangents to these circles, the values of c' and φ' for each soil layer are listed in Table 4.5.

Fig. 4.6 ε_1/q-ε_1 curves

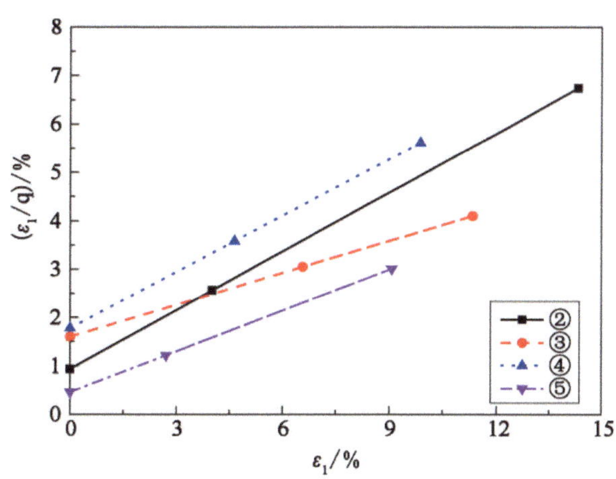

Table 4.4 Values of q_a and R_a of soil layers(②–⑤)

	q_f/kPa	q_a/kPa	R_f
②	223.7	246.7	0.91
③	307.6	454.3	0.68
④	185.0	257.7	0.72
⑤	317.3	355.5	0.89

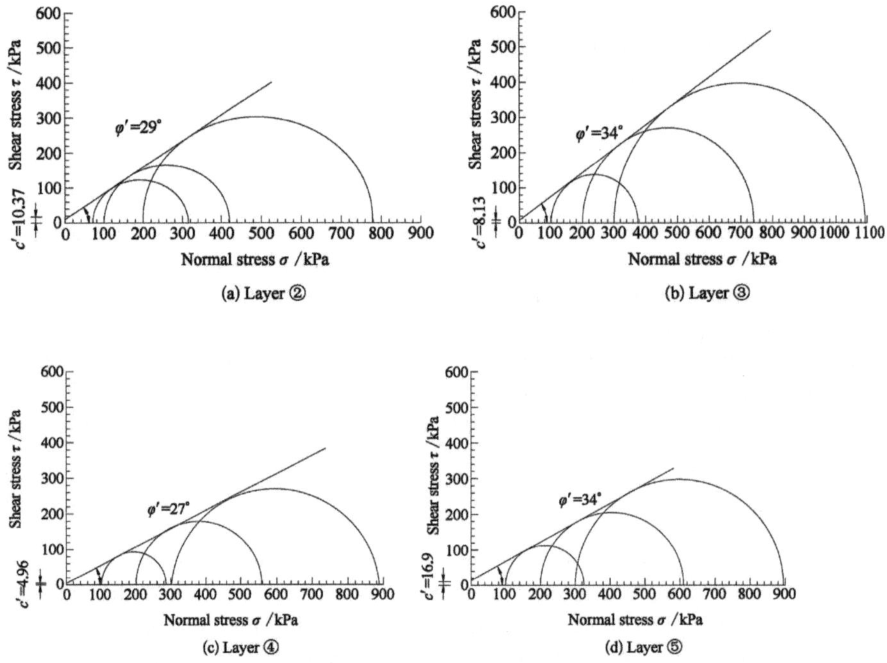

Fig. 4.7 Mohr circle of soil layers (②–⑤)

Table 4.5 Values of c', φ' of soil layers (②–⑤)

	②	③	④	⑤
c'	10.37	8.13	4.96	16.90
φ'	29	34	27	34

(2) Results of the triaxial consolidated drained loading–unloading shear test

To obtain the reference loading–unloading modulus E_{ur}^{ref}, triaxial loading–unloading shear tests are conducted on soil samples from layers ②–⑤, using a reference confining pressure p^{ref} of 100 kPa. The curves between the deviatoric stress and axial strain for each soil layer are shown in Fig. 4.8.

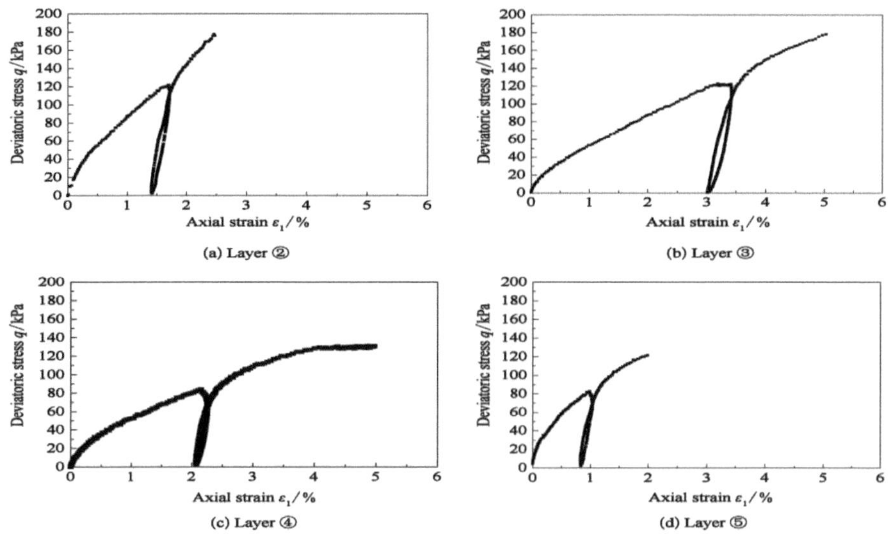

Fig. 4.8 Strain–stress curves of triaxial CD loading–unloading tests on soil layers (②–⑤)

It can be observed from Fig. 4.8 that the axial strain slightly increases at the initial stage of unloading. When unloading reaches a certain level, the axial strain slightly decreases. Overall, this pattern indicates a certain degree of rebound. During the reloading process, the initial stress–strain curve is very steep, while later it becomes much flatter. The test curves for soil samples from each layer during the loading–unloading process exhibit a hysteresis loop. By connecting the two endpoints of the hysteresis loop, the slope of this line is expressed as E_{ur}^{ref}. The values E_{ur}^{ref} for the soil samples from layers ②–⑤ are 38.8, 33.9, 36.6, and 42.4 MPa respectively.

(3) Results of the standard consolidation test

The curves of axial load versus axial strain for each soil layer are shown in Fig. 4.9, where the curve fitting for the tests of each layer have an R^2 value of 0.99. As can be seen from Fig. 4.9, the trend of axial strain variation with axial load for the four soil samples is similar. The initial part of the curve is relatively flat, but as the axial load increases, the slope of the curve becomes steeper as well. Based on the curve fitting for each layer, the slope of the curve at an axial load of 100 kPa can be obtained, which represents the reference tangent modulus E_{oed}^{ref}.

According to the proportionate relationship between the void ratio and axial strain, the relationship curve between the axial load p and the void ratio e_i can be obtained, as shown in Fig. 4.10. From Fig. 4.10, it is observed that the initial void ratios of soil samples from layers ③ and ④ are greater than those of layers ② and ⑤. As the axial load increases, the void ratio of each layer decreases. The change in void ratio for layer ③ is significantly greater than that of the other three layers; the changes in void ratio for layers ② and ⑤ are relatively small. At the same time, the compression modules for soil samples from layers ②–⑤ are obtained E_{s12}, as shown in Table 4.6.

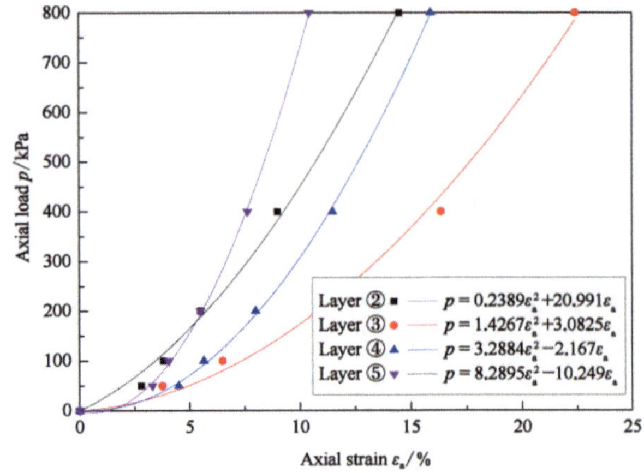

Fig. 4.9 Relationship between load and strain of consolidation tests

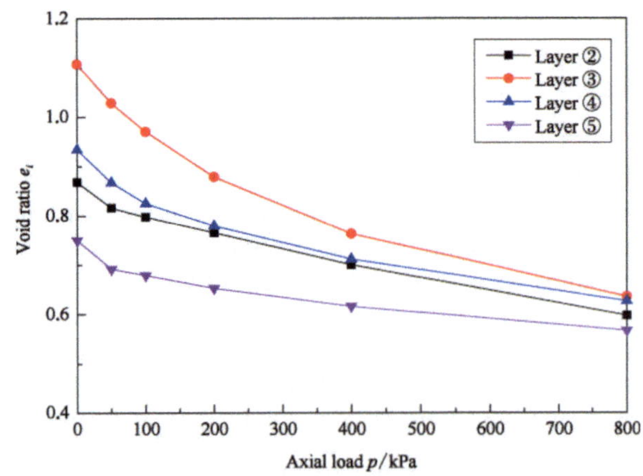

Fig. 4.10 Relationships between load and void ratio of layers in consolidation tests

Table 4.6 Values of E_{oed}^{ref} and E_{s12} of soil layers (②–⑤) (unit: MPa)

	Layer ② soil	Layer ③ soil	Layer ④ soil	Layer ⑤ soil
$E_{\text{oed}}^{\text{ref}}$	3.7	2.5	3.6	5.9
E_{s12}	5.9	2.3	4.3	6.7

(4) Results of the resonant column test

To obtain the small-strain stiffness values of typical soil layers in Shanghai, the dynamic shear modulus G_d under various dynamic shear strains γ_d for Shanghai soil layers ②–⑤ are measured using the V. P. Drnevich resonant column apparatus. The confining pressures for soil samples from layers ②–⑤ are set based on the vertical effective overburden stress at their respective sampling depths, namely 30, 70, 150, and 200 kPa. According to the analysis by Hardin B. O. et al., the characteristics of the $G_d - \gamma_d$ curve can be described by a hyperbolic model, that is, $1/G_d = a + b$, where a and b are experimental constants. These constants can be calculated through regression statistical analysis of the experimental data. As γ_d approaches zero and $1/G_d$ approaches a, G_d is represented by G_0, where $G_0 = 1/a$. G_0 is referred to as the initial shear modulus. The fitting curves for the four soil layers and their initial shear modulus are presented in Table 4.7.

The value $\gamma_{0.7}$ denotes the shear strain corresponding to the point at which secant shear modulus decays to 70% of the initial shear modulus. According to Fig. 4.11, the $\gamma_{0.7}$ values for the soil samples from layers ②–⑤ are 2.9×10^{-4}, 2.5×10^{-4}, 2.7×10^{-4} and 3.4×10^{-4}.

According to the formula in the PLAXIS software manual:

$$G_0 = G_0^{\text{ref}} \left(\frac{c' \cos\varphi' - \sigma_3' \sin\varphi'}{c' \cos\varphi' + p^{\text{ref}} \sin\varphi'} \right)^m \tag{4.27}$$

The reference initial shear modules G_0^{ref} of the various layers can be obtained under the reference confining pressure $p^{\text{ref}} = 100$ kPa, as presented in Table 4.8.

(4) Comparative analysis of experimental results

Through the four parts of the experiment, the soil parameters obtained for Shanghai layers ②–⑤ include the effective strength parameters c' and φ'; the modulus values

Table 4.7 Values of fitting parameters and G_0 of soil layers (②–⑤)

		Layer ② soil	Layer ③ soil	Layer ④ soil	Layer ⑤ soil
Confining pressure		30	70	150	200
Curve fitting	a/MPa^{-1}	0.045	0.034	0.026	0.018
	b/MPa^{-1}	68.29	60.63	41.05	25.42
G_0/MPa		22.1	29.2	38.6	53.8

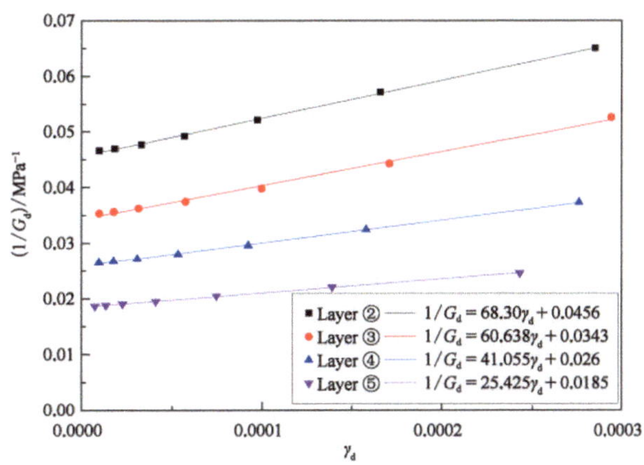

Fig. 4.11 Curves of shear modulus and strain of soil layers

Table 4.8 Values of G_0^{ref} of soil layers (②–⑤)

	Layer ② soil	Layer ③ soil	Layer ④ soil	Layer ⑤ soil
p^{ref}/kPa	100	100	100	100
G_0^{ref}/MPa	45.0	37.4	28.6	33.6

E_{50}^{ref}, E_{ur}^{ref} and E_{oed}^{ref} at a reference stress p^{ref} of 100 kPa and the failure ratio R_f; the reference small-strain parameters G_0^{ref} and $\gamma_{0.7}$. The values of these parameters are summarized and analyzed as follows:

(1) Comparative analysis of values of E_{50}^{ref}, E_{ur}^{ref}, E_{oed}^{ref} and R_f

The parameters including values of E_{50}^{ref},E_{ur}^{ref}, E_{oed}^{ref}, R_f obtained from the tests for the typical soil layers in the Shanghai area are summarized in Table 4.9. For comparison, the table also lists the parameters of the soil hardening model for Shanghai and other regions obtained from Reference (Wang et al. 2012). It can be seen from Table 4.9 that the values of c', E_{50}^{ref},E_{ur}^{ref},E_{oed}^{ref}, E_{s12} and R_f in the layers ③ and ④ in the Shanghai area are lower than those for layers ② and ⑤.

The ratio of E_{50}^{ref} and E_{oed}^{ref} for the four soil samples in the tests is essentially the same [$E_{50}^{ref} = (1.08–1.2)E_{oed}^{ref}$], which is similar to the findings in Reference (Wang et al. 2012) [$E_{50}^{ref}= (0.9–1.3)E_{oed}^{ref}$]. The ratio [$E_{ur}^{ref}= (6.7–8.4)E_{50}^{ref}$] of E_{ur}^{ref} and E_{50}^{ref} for soil layers ② and ⑤ is relatively close. The ratio [$E_{ur}^{ref}= (9.4–11.6)E_{50}^{ref}$] for soil layers ③ and ④ is also relatively close. The values of E_{oed}^{ref} are 0.63–1.06 times the values of E_{s12} for soil layers ②–⑤ in the tests, which is quite close to the results [$E_{oed}^{ref}=(0.9–1.1)E_{s12}$] in Reference (Wang et al. 2012).

Table 4.9 also covers some parameters of soils from other areas. The effective cohesion *c'* of the soils from the Shanghai area obtained in the tests is generally higher than that of soils from other areas, while the effective angle of internal friction *φ'* is similar to that of other areas. The failure ratio R_f values for layers ③ and ④ in the Shanghai area are lower than those of other areas, whereas the R_f values for layers ② and ⑤ are comparable to those elsewhere. The E_{50}^{ref} values for the Shanghai area are 1.08–1.2 times the E_{oed}^{ref} values. This is comparable to 1.7 times for silty clay in the Wuxi area, 0.5–1.1 times for soft soil in Binhai, Tianjin, and 1.0 times for the Beijing area. The E_{ur}^{ref} values for the Shanghai area soils are 6.7–11.6 times the E_{50}^{ref} values, which is a higher ratio than that of soils from other regions. The E_{oed}^{ref} values for Shanghai are 0.63–1.06 times the E_{s12} values, which is relatively close to 1.1 times for the Beijing area but lower than 1.4–2.1 times for Binhai, Tianjin.

(2) Comparative analysis of small-strain parameters

(1) Comparative analysis of values of G_0^{ref}

Table 4.10 lists the G_0^{ref} values and the E_{ur}^{ref} values for each soil layer and provides the proportional relationship between G_0^{ref} and E_{ur}^{ref} for each layer.

From Table 4.10, it can be seen that the layer ② has the highest G_0^{ref} value, followed by layers ③ and ⑤, with layer ④ having the lowest value. The ratio of G_0^{ref} to the E_{ur}^{ref}

Table 4.9 Comparisons of parameters of HSS model

Soil name		c/kPa	φ'/(°)	E_{oed}^{ref}/MPa	E_{50}^{ref}/MPa	E_{ur}^{ref}/MPa	R_f	E_{s12}/MPa	Proportional relationship of modulus
Shanghai area	② Clay	10.37	29	3.7	4.7	38.8	0.91	5.9	$E_{50}^{ref} = 1.2E_{oed}^{ref}$ $E_{ur}^{ref} = 84E_{50}^{ref}$ $E_{oed}^{ref} = 0.63E_{s12}$
	③ Muddy silty clay	8.13	34	2.5	3.0	33.9	0.68	2.3	$E_{50}^{ref} = 1.2E_{oed}^{ref}$ $E_{ur}^{ref} = 11.6E_{50}^{ref}$ $E_{oed}^{ref} = 1.06E_{s12}$
	④ Muddy clay	4.96	27	3.6	3.9	36.6	0.72	4.3	$E_{50}^{ref} = 1.08E_{oed}^{ref}$ $E_{ur}^{ref} = 9.4E_{50}^{ref}$ $E_{oed}^{ref} = 0.85E_{s12}$
	⑤ Silty clay	16.90	34	5.9	6.4	42.4	0.89	6.7	$E_{50}^{ref} = 1.08E_{oed}^{ref}$ $E_{ur}^{ref} = 6.7E_{50}^{ref}$ $E_{oed}^{ref} = 0.87E_{s12}$
Shanghai area (Wang et al..2012)	② Clay	10	25.2	3.4	4.4	19.3	0.96	3.9	$E_{50}^{ref} = 1.3E_{oed}^{ref}$ $E_{ur}^{ref} = 4.4E_{50}^{ref}$ $E_{oed}^{ref} = 0.9E_{s12}$

(continued)

Table 4.9 (continued)

Soil name		c/kPa	φ'/(°)	E_{oed}^{ref}/MPa	E_{50}^{ref}/MPa	E_{ur}^{ref}/MPa	R_f	E_{s12}/MPa	Proportional relationship of modulus
	③ Muddy silty clay	10	20.1	1.2	1.6	14.9	0.58	1.3	$E_{50}^{ref} = 13E_{oed}^{ref}$ $E_{ur}^{ref} = 9.3E_{50}^{ref}$ $E_{oed}^{ref} = 1.1E_{s12}$
	④ Muddy clay	3	27.3	1.9	2.0	15.6	0.54	2.2	$E_{50}^{ref} = 1.1E_{oed}^{ref}$ $E_{ur}^{ref} = 7.8E_{50}^{ref}$ $E_{oed}^{ref} = 0.9E_{s12}$
	⑤ Silty clay	20	26.7	6.1	5.5	23.5	0.95	6.3	$E_{50}^{ref} = 0.9E_{oed}^{ref}$ $E_{ur}^{ref} = 4.3E_{50}^{ref}$ $E_{oed}^{ref} = E_{s12}$
Wuxi area (Chen 2011)	③$_2$ Silty clay			4.8	8.0	25.2			$E_{50}^{ref} = 0.7E_{oed}^{ref}$ $E_{ur}^{ref} = 3.2E_{50}^{ref}$
Tianjin Binhai soft soil area (Liu 2008)	② Silty clay			3.8	2.8			1.8	$E_{50}^{ref} = 0.7E_{oed}^{ref}$ $E_{oed}^{ref} = 2.1E_{s12}$
	③ Muddy silty clay			3.5	2.3			1.8	$E_{50}^{ref} = 1.7E_{oed}^{ref}$ $E_{oed}^{ref} = 2.0E_{s12}$
	④ Clay			5.0	5.3			2.8	$E_{50}^{ref} = 1.1E_{oed}^{ref}$ $E_{oed}^{ref} = 1.8E_{s12}$

(continued)

Table 4.9 (continued)

Soil name		c/kPa	φ'/(°)	E_{oed}^{ref}/MPa	E_{50}^{ref}/MPa	E_{ur}^{ref}/MPa	R_f	E_{s12}/MPa	Proportional relationship of modulus
	⑤ Silty clay			5.6	2.9			4.0	$E_{50}^{ref} = 0.5E_{oed}^{ref}$ $E_{oed}^{ref} = 1.4E_{s12}$
Beijing area (Zhou 2010)	Silty clay			4.0	3.9	17.8	0.8	3.6	$E_{50}^{ref} = E_{oed}^{ref}$ $E_{ur}^{ref} = 4.6E_{50}^{ref}$ $E_{oed}^{ref} = 1.1E_{s12}$

Table 4.10 Values of G_0^{ref} and E_{ur}^{ref} of soil layers (②–⑤)

Soil layers	E_{ur}^{ref}/MPa	G_0^{ref}/MPa	Proportional relationships
② Clay	38.8	45.0	G_0^{ref}=1.2 E_{ur}^{ref}
③ Muddy silty clay	33.9	37.4	G_0^{ref}=1.1 E_{ur}^{ref}
④ Muddy clay	36.6	28.6	G_0^{ref}=0.8 E_{ur}^{ref}
⑤ Silty clay	42.4	33.6	G_0^{ref}=0.8 E_{ur}^{ref}

for each layer is relatively similar, $G_0^{ref} = (0.8-1.2)E_{ur}^{ref}$. Bao et al. (1989) used a STOKOE torsional resonant column apparatus to test Shanghai gray soft clay (buried at 19.6–21.6 m depth) under a confining pressure of 100 kPa, obtaining an initial shear modulus G_0 of approximately 30 MPa. The value is slightly lower than the reference initial shear modulus value of 33.6 MPa obtained for the silty clay layer of Shanghai's layer ⑤ in the tests. Jiang (2009) employs the CKC instrument and a local measurement system to acquire Shanghai's silty clay at a depth of 23.3 m. Under a confining pressure of 186 kPa and a shear strain of 2×10^{-5}, the shear stiffness value reaches approximately 62.5 MPa. Calculations have shown that at a confining pressure of 100 kPa and a shear strain of 2×10^{-5}, the shear stiffness value is about 43 MPa, slightly higher than the reference initial shear modulus value of 33.6 MPa obtained for the silty clay layer of Shanghai's layer ⑤ in the tests. Thus, the reference initial shear modulus values obtained from the tests in this section fall between the results of References (Bao et al. 1989; Jiang 2009), with all three values showing relatively minor differences.

② Comparative analysis of values of $\gamma_{0.7}$

The studies of Vucetic and Dobry (1991) and Stokoe et al. (2004) indicate that the value of $\gamma_{0.7}$ of clay is significantly influenced by the soil's plasticity index (I_P) and the overconsolidation ratio (*OCR*). Stokoe provides a formula for calculating the value of $\gamma_{0.7}$ of clay:

$$\gamma_{0.7} = (\gamma_{0.7})_{ref} + 5 \times 10^{-6} I_P (ORC)^{0.3} \tag{4.28}$$

In the equation: I_P stands for the plasticity index; *OCR* represents the overconsolidation ratio of the soil; $(\gamma_{0.7})_{ref}$ denotes the shear strain in terms of $I_P = 0$.

Wang et al. (2013) computes and provides an approximate range for the value of $\gamma_{0.7}$ of typical clay layers in Shanghai, which is highlighted in Table 4.11. This table also includes the experimental values obtained in this section. Upon comparison, it is acknowledged that the value of $\gamma_{0.7}$ obtained from the tests align closely with the empirical calculations.

(3) Summary of model parameters

This section has obtained 8 parameters of the HSS model for soil layers ②–⑤ in Shanghai through laboratory tests: *c'*, *φ'*, E_{oed}^{ref}, E_{50}^{ref}, E_{ur}^{ref}, R_f, G_0^{ref} and $\gamma_{0.7}$, along

Table 4.11 Values of 0.7 of soil layers (②–⑤)

Soil layers	$\gamma_{0.7}$(The tests)/($\times 10^{-4}$)	$\gamma_{0.7}$ (Reference Wang et al. 2013)/($\times 10^{-4}$)
② Clay	2.9	2.0–2.8
③ Muddy silty clay	2.5	1.5–1.9
④ Muddy clay	2.7	1.9–2.3
⑤ Silty clay	3.4	1.6–2.3

with their proportional relationships, as shown in Table 4.12. For actual engineering projects, field investigation reports are generally provided with the values of c', φ' and E_{s12} of the soil. Therefore, the parameters including c', φ', E_{50}^{ref}, E_{ur}^{ref}, E_{oed}^{ref}, R_f, G_0^{ref} and G_0^{ref} can be determined based on the investigation report and the proportional relationships listed in Table 4.12. The parameters K_0, ψ, p^{ref}, ν_{ur} and m can be taken from the values in Table 4.2.

In summary, the HSS model consists of a total of 13 parameters. The definitions and values of the parameters can be known in Table 4.13.

(5) Project verification

To verify the suitability of the HSS constitutive model for deep excavation projects in the Shanghai area, this section focuses on a deep pit excavation project in Pudong, Shanghai. The excavation process was simulated using Zsoil software for full 3D modeling, and the computational results were compared with field monitoring data (Liang et al. 2017).

(1) Project overview

The project is located in the Lujiazui Financial and Trade Zone in the Pudong New District of Shanghai, with the foundation pit depths ranging from 14.75 to 22.8 m. The surrounding conditions are complex. The south side of the site is closely adjacent to Metro Line 4, and Metro Line 6 runs across the entire site as an open-cut underground structure at B1 level.

Table 4.12 Prime parameters of HSS model of soil layers (②–⑤)

Soil layers	E_{s12}/MPa	E_{oed}^{ref}/MPa	E_{50}^{ref}/MPa	E_{ur}^{ref}/MPa	G_0^{ref}/MPa	γ_{07}/($\times 10^{-4}$)	R_f	c'	φ'
② Clay	5.9	3.7	4.7	38.8	45.0	2.9	0.91	10.37	29
		$0.63E_{s12}$	$1.2E_{oed}^{ref}$	$8.4E_{50}^{ref}$	$1.2E_{ur}^{ref}$				
③ Muddy silty clay	2.3	2.5	3.0	33.9	37.4	2.5	0.68	8.13	34
		$1.06E_{s12}$	$1.2E_{oed}^{ref}$	$11.6E_{50}^{ref}$	$1.1E_{ur}^{ref}$				
④ Muddy clay	4.3	3.6	3.9	36.6	28.6	2.7	0.72	4.96	27
		$0.85E_{s12}$	$1.08E_{oed}^{ref}$	$9.4E_{50}^{ref}$	$0.8E_{ur}^{ref}$				
⑤ Silty clay	6.7	5.9	6.4	42.4	33.6	3.4	0.89	16.9	34
		$0.87E_{s12}$	$1.08E_{oed}^{ref}$	$6.7E_{50}^{ref}$	$0.8E_{ur}^{ref}$				

Table 4.13 The definitions of the HSS model parameters and evaluation of parameters of the HSS model for Shanghai soils

Definitions of parameters	Empirical values of Shanghai soils
C': effective cohesion of soil	Conventional triaxial consolidated drained tests (investigation report-based)
Φ': effective internal friction angle of soil	Conventional triaxial consolidated drained tests (investigation report-based)
K_0: coefficient of lateral earth pressure "at rest" of normally-consolidated soil	Cohesive soil: 0.95-sin*φ'*; sandy and silty soil: 1-sin*ϕ'*
Ψ: dilatancy angle of soil	For sandy soils, $\psi = \varphi' - 30°$, and when *φ'* is less than 30°, *ψ* is set to 0; for cohesive soils, *ψ* is set to 0
m: exponential power of stiffness stress level	For sandy and silty soil, a value of 0.5 is generally used; for cohesive soil, the value ranges from 0.5 to 1.0
p^{ref}: reference stress	100 kPa
ν_{ur}: loading–unloading Poisson's ratio	0.2
$E_{\text{oed}}^{\text{ref}}$: reference tangent modulus for consolidation tests	$E_{\text{oed}}^{\text{ref}} \approx (0.63\text{–}1.06)E_{s12}$ (E_{s12} is based on investigation reports)
E_{50}^{ref}: reference secant modules from drained triaxial tests	$E_{50}^{\text{ref}} \approx 1.08\text{–}1.2E_{\text{oed}}^{\text{ref}}$
$E_{\text{ur}}^{\text{ref}}$: reference loading–unloading modules from drained triaxial tests	$E_{\text{ur}}^{\text{ref}} \approx (6.7\text{–}8.4)E_{50}^{\text{ref}}$
R_f: failure ratio	0.5–0.9
G_0^{ref}: reference initial shear modules	$G_0^{\text{ref}} \approx (0.8\text{–}1.2)E_{\text{ur}}^{\text{ref}}$
$\gamma_{0.7}$: the shear strain corresponding to the point at which secant shear modulus decays to 70% of the initial shear modulus	$(2.0\text{–}5.0) \times 10^{-4}$

The site is divided by Line 6 into two main excavation zones (A1 and B1) and 15 smaller symmetrical zones along both sides of Line 6 and the north side of Line 4 (A2-A8, B2-B9). The excavation sequence starts with the larger pits, B1 and A1, followed by the symmetrical excavation of the smaller pits on both sides. The sketch diagram of location of pits is illustrated in Fig. 4.12.

This project uses the open-cut and building method for construction. The excavation depth of the large pits is about 22.8 m, with a diaphragm wall thickness of 1 m. Within the plane of the foundation pit, a comprehensive cross-bracing system is utilized with five levels of C40 reinforced concrete supports (elastic modulus E = 32.5GPa). The first level of support measures 1200 mm by 1000 mm, while the second to fifth levels are 1400 mm by 1000 mm, with a horizontal spacing of about 9 m. The smaller pits, A2–A4 and B2–B4, are excavated to approximately 19.55 m and equipped with five levels of support. The first level consists of reinforced concrete supports, while the remaining four levels use *Φ*609 steel pipes (t = 16 mm). The smaller pits, A5–A8 and B5–B8, which are excavated to about 14.75 m, four levels of

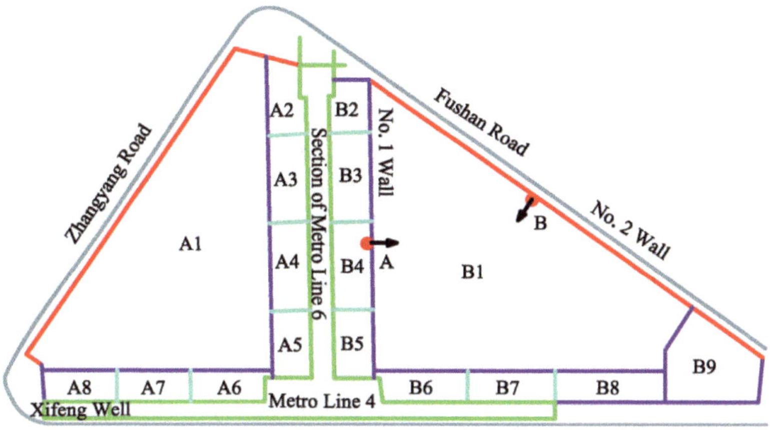

Fig. 4.12 Sketch diagram of location of pits

supports are arranged. The first level consists of reinforced concrete supports, while the remaining three levels use Φ609 steel pipes ($t = 16$ mm). Adjacent to the large pit of Line 6, the diaphragm wall has a thickness of 1.0 m and a depth of 40 m, while the other large pits have diaphragm walls of 1.0 m in thickness and 50 m in depth. All smaller pits have diaphragm walls that are 0.8 m thick and 40 m deep.

(2) Calculation model and parameter evaluation

Numerical analysis is performed using the Zsoil finite element software. The soil-structure model is conducted employing the HSS constitutive model. The HSS model is based on the HS model, which is an isotropic hardening elasto-plastic model that incorporates a cap yield surface. The shearing yield surface follows a non-associated flow rule. The HSS model further considers the variation characteristics of soil stiffness within the small strain range. The computational model spans 1000 m by 500 m by 100 m (length × width × depth), with model boundaries all situated more than five times the maximum excavation depth away from the foundation pit. The entire numerical model includes 46,170 spatial hexahedral 8-node solid elements (for soil simulation), 15,749 Beam elements (for enclosure and internal supports simulation), 1,680 Construct continuum elements (for tunnel structure simulation), 1,680 Shell elements (for diaphragm wall and excavation base slab simulation), and 325 Pile elements (for column piles simulation).

Evaluation of soil parameters:

For layers ②–⑤, parameters including c', φ', $E_{\text{oed}}^{\text{ref}}$, E_{50}^{ref}, $E_{\text{ur}}^{\text{ref}}$, R_{f}, G_0^{ref} and $\gamma_{0.7}$ are taken from experimental results as shown in Table 4.11. The values for others are taken as $\psi = 0$, $p^{\text{ref}} = 100$, $m = 0.8$, $\nu_{\text{ur}} = 0.2$. K_0 values are set at 0.46, 0.50, 0.56, and 0.49 respectively, as per the reference in Table 4.2.

For layers ⑥–⑨, basic parameters such as c', φ', K_0, unit weight γ, Poisson's ratio ν, and compression modulus E_{s12} come from the engineering investigation report. The dilatancy angle ψ is referenced from Table 4.2.

Other parameters for layers ⑥–⑨ are as follows: Layer ⑥ is a typical clay layer, and its values for parameters such as E_{oed}^{ref}, E_{50}^{ref}, E_{ur}^{ref}, R_f, G_0^{ref} and $\gamma_{0.7}$ can be taken as the average values of layer ② and ⑤, which are also clay layers: $E_{50}^{ref} = 1.14E_{oed}^{ref}$, $E_{ur}^{ref} = 7.6E_{50}^{ref}$, $G_0^{ref} = E_{ur}^{ref}$, $E_{oed}^{ref} = 0.85E_{s12}$, $R_f = 0.9$, $\gamma_{0.7} = 2.8 \times 10^{-4}$.

Since layers ⑦–⑨ are sandy soil layers, there are no reported experimental studies on E_{oed}^{ref}, E_{50}^{ref} and E_{ur}^{ref} for sandy soil layers in Shanghai. Wang (2012) has collected 13 numerical analysis examples using the HSS or HS mode. It is found that E_{50}^{ref} of sandy soil is 0.7–1.25 times than that of E_{oed}^{ref} and E_{ur}^{ref} is 3–5.6 times than that of E_{50}^{ref}. Therefore, for sandy soil layers ⑦–⑨, it is reasonable to take $E_{50}^{ref} = E_{oed}^{ref}$, $E_{ur}^{ref} = 4E_{50}^{ref}$. Similarly, following the practice for clay layers, $G_0^{ref} = E_{ur}^{ref}$, $E_{oed}^{ref} = 0.85E_{s12}$.

Thus, the parameters of the HSS model of soil layers ⑥–⑨ at the project site can be obtained, as shown in Table 4.14.

(3) Analysis of computational results

Figure 4.13 illustrates points A and B on Wall 1 and Wall 2, respectively, to analyze the deformation characteristics of the retaining walls along their depth. The figure presents a comparison between actual measurements and numerical calculations of the wall deformations during the excavation of the large pit B1. As the excavation of pit B1 progresses, the horizontal displacement of No.1 and No.2 Walls gradually increases, with the position of maximum displacement moving downwards. Upon reaching the bottom of pit B1, the calculated maximum deformation for No.1 Wall is 47.1 mm, approximately 0.207% of H, with an actual measurement of 48.4 mm, approximately 0.212% of H. In terms of No.2 Wall, the calculated maximum deformation is 52.2 mm, approximately 0.23% of H, with an actual measurement of 58.8 mm, approximately 0.26% of H (H represents the excavation depth of the pit). The finite element calculations align well with the actual measurements, indicating a good match in the pattern of wall deformation.

Table 4.14 Parameters of the HSS model of soil layers (⑥–⑨)

Layer	c'	φ'	ψ	p^{ref}/kPa	m	ν_{ur}	G_0^{ref}/MPa	E_{oed}^{ref}/MPa	E_{50}^{ref}/MPa	E_{ur}^{ref}/MPa	K_0	R_f	$\gamma_{0.7}$/($\times 10^{-4}$)	γ/(kN/m^3)
⑥	5	33	0	100	0.8	0.2	52.6	6.1	6.9	52.6	0.46	0.9	2.8	19.7
⑦	0	34.5	4.5	100	0.5	0.2	42.4	10.6	10.6	42.4	0.37	0.9	2.8	18.8
⑧	0	35	5	100	0.5	0.2	51.2	12.8	12.8	51.2	0.34	0.9	2.8	19.0
⑨	0	35.1	5.1	100	0.5	0.2	50.3	12.6	12.6	50.3	0.34	0.9	2.8	19.5

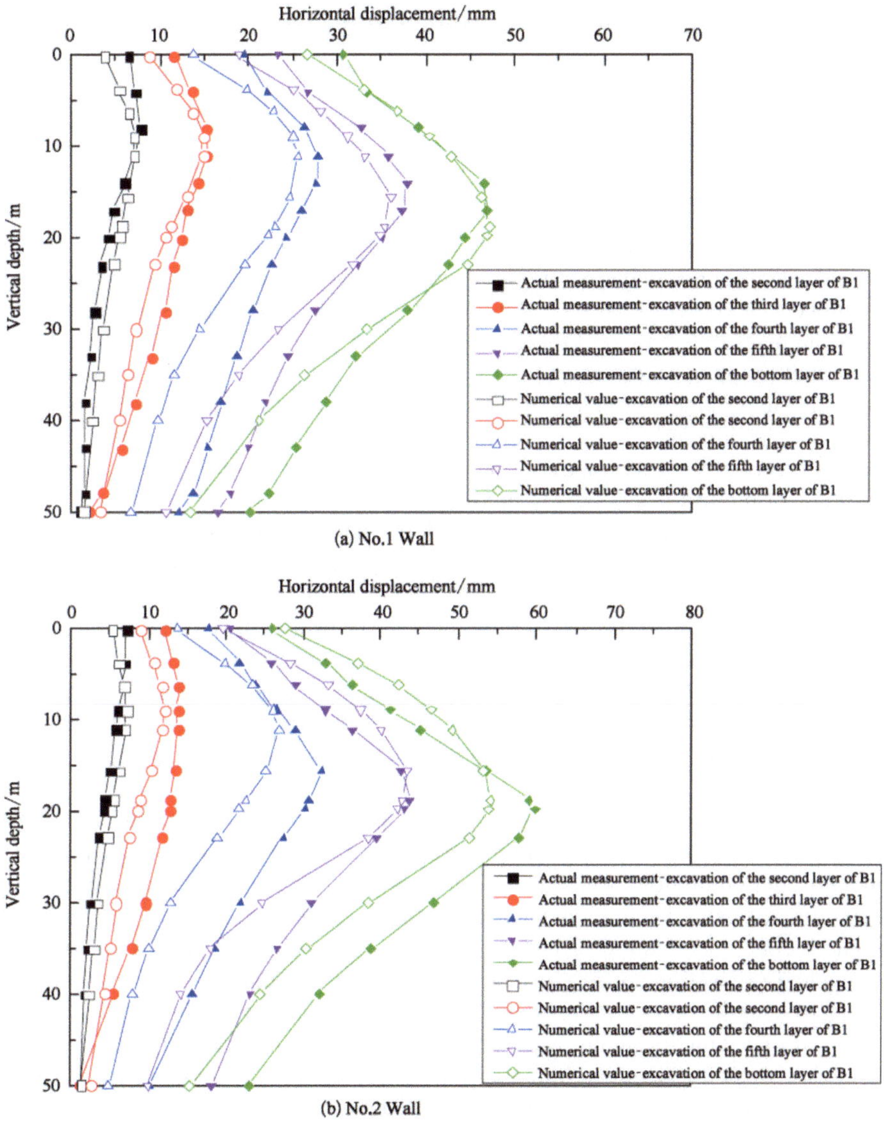

Fig. 4.13 Comparisons of horizontal displacement of diaphragm walls

4.3 Numerical Calculation Models and Working Condition Settings

4.3.1 The Finite Element Model

Considering the impact of excavation of pits of different scales on underlying tunnels, the dimensions of the finite element model boundary are set as 150.0 m × 100.0 m. Considering the common medium-deep tunnels, the thickness of the soil layer above the tunnel, H_0, is set to 15.0 m. The outer diameter of the tunnel is 6.2 m, with an inner diameter of 5.5 m, and the lining wall thickness is 0.35 m. The structure of the tunnel lining is as shown in Fig. 3.2. The soil layer is chosen from the ancient river channel sedimentation area in Shanghai, with all foundation soils belonging to Quaternary sedimentary deposits composed of clay, silty soil, and sandy soil. Taking the soil layers of a certain metro monitoring project in Shanghai as an example, the overview of the soil layers is presented in Table 4.15.

The calculation employs a two-dimensional plane strain model to simulate the excavation and support form of a pit with diaphragm walls and horizontal bracing. The pit enclosure structure uses a diaphragm wall with a thickness of 0.8 m and an insertion ratio of 1:1.5. The first horizontal brace is installed 1.0 m below the ground surface, with the remaining spaces ranging from 4.5 to 5.0 m. The model configuration includes: a free boundary at the top surface, normal displacement constraints at lateral boundaries, and full displacement constraints at the base. The calculation employs fluid–solid coupling analysis, with the water level dropping 1.0 m the excavation surface before excavation.

The simulation of the soil elements is conducted using solid elements, with mesh division considering soil layers and pit internal supports. Structures such as the pit's underground continuous wall, braces, bottom slab, bored piles, and tunnel lining are simulated using beam elements. According to the excavation area and research requirements, construction processes such as uplift piles, soil reinforcement, and segmented excavation are set up. The finite element calculation model is shown in Fig. 4.14, and the tunnel lining beam element model is shown in Fig. 4.15.

Given that tunnel segment rings are not completely uniform stiffness rings, it's formed of six bolt-assembled linings, whose stiffness at the joints is reduced. Therefore, the force performance of the segments is no longer an isolated structural issue but a soil-structure interaction problem. The presence of joints in the lining ring affects the overall stress response of segments. Joint displacement (such as opening or rotation) trigger a redistribution of internal forces and affects the overall deformation of the segments. Likewise, the interaction between the ground and the structure affects the internal forces. The impact of the ground on the segments cannot be overlooked. The segment joints are influenced by various factors such as assembly forces, longitudinal thrust, and sealing materials, exhibiting nonlinear characteristics. Therefore, the simulation analysis of tunnel segments must take into account the impact of joints.

Table 4.15 Overview of the soil layers

Geologic time scale	No. of soil layer	Name of soil layer	Formation type	Thickness of layer/m	Elevation at layer base/m	Description
Q_4^3	①$_1$	Miscellaneous fill	Anthropogenic factors	1.50–4.90 2.50	2.44 to −1.08 1.54	The upper part is mainly construction waste, followed by clayey soil, containing concrete blocks, bricks, etc., with numerous plant roots, resulting in uneven soil quality
Q_4^2	②	Brownish-yellow to grayish-yellow silty clay	Binhai–shallow sea	0.60–2.40 1.42	0.17 to −0.46 −0.05	Contains iron oxide rust spots and iron-manganese nodules, with a small number of thin layers of silt. The soil quality is relatively uniform
	③	Gray muddy silty clay	Binhai–shallow sea	3.10–4.80 3.94	−2.55 to −5.48 −3.77	Contains mica and organic matter, with numerous thin layers of silt, resulting in uneven soil quality
Q_4^1	④	Gray muddy clay	Binhai, swamp	5.40–9.10 7.60	−9.54 to −13.21 −11.37	Contains mica and organic matter, with a small number of thin layers of silty sand. The soil quality is uniform
	⑤$_{11}$	Gray clay	Binhai, swamp	3.00–9.00 5.88	−14.71 to −21.12 −17.25	Contains mica and spots of organic matter, with mud, calcareous nodules, and semi-humified roots. The soil quality is uniform
	⑤$_{12}$	Gray silty clay	Binhai, swamp	1.40–12.00 6.39	−17.40 to −27.23 −23.79	Contains mica and spots of organic matter, with occasional grey-white mud calcareous nodules. The soil quality is relatively uniform
	⑤$_{31}$	Gray silty clay mixed with silty soil	Palaeochannel	5.00–19.60 9.87	−30.08 to −36.21 −32.84	Contains mica, with areas predominantly composed of clayey silt, resulting in uneven soil quality
	⑤$_{32}$	Gray silty clay	Palaeochannel	2.10–10.50 5.23	−35.80 to −46.71 −38.07	Contains mica and organic matter, with thin layers of silt. The soil quality is relatively uniform

(continued)

Table 4.15 (continued)

Geologic time scale	No. of soil layer	Name of soil layer	Formation type	Thickness of layer/m	Elevation at layer base/m	Description
Q_4^1	⑤$_4$	Grayish-green silty clay	Palaeochannel	1.50–6.50 3.32	−38.71 to −46.09 −41.33	Contains iron oxide and spots of organic matter, with a few thin layers of silt. The soil quality is relatively uniform
Q_3^2	⑦$_1$	Grayish-yellow to gray sandy silty soil	Estuary–Binhai	0.90–6.00 3.08	−40.96 to −47.18 −44.40	Contains mica, with numerous thin layers of silt, resulting in uneven soil quality
	⑦$_{21}$	Grass-yellow to gray silty sand	Estuary–Binhai	4.00–16.60 12.21	−44.58 to −58.31 −55.22	Composed of mineral particles such as feldspar, quartz, and mica, with thin layers of clayey soil. The sandy quality is relatively pure
Q_3^1	⑨$_1$	Gray silty sand	Binhai–Estuary	18.50–22.90 20.19	−81.31 to −85.30 −83.30	Composed of mineral particles such as feldspar, quartz, and mica, with local areas of sandy silt. The sandy quality is relatively pure

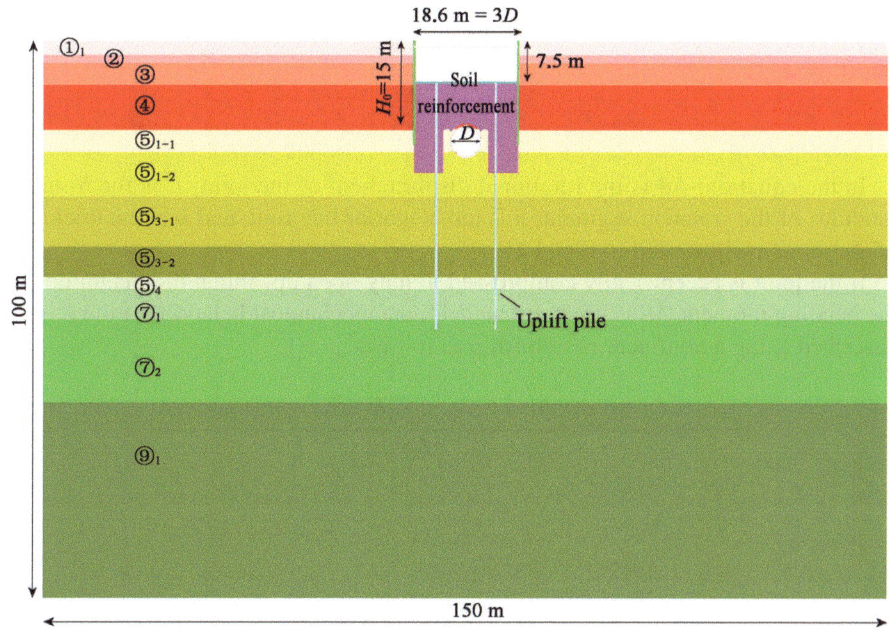

Fig. 4.14 The finite element calculation model

Fig. 4.15 Tunnel segment beam element model and hinge joint model (Beam-Hinge Joint Model)

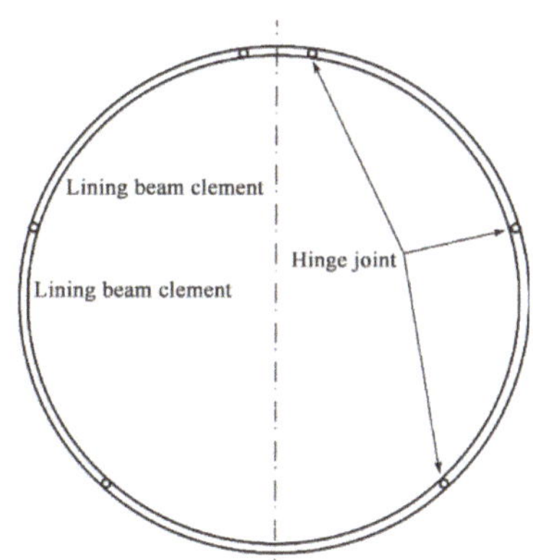

Janssen proposed a simple theoretical model to describe the mechanical behavior of tunnel segment joints (Janssen 1983). The model assumes that the joint can be equivalently represented as a linear elastic concrete beam in full contact with the lining surface (Fig. 4.16). Under loading, the segment joint can rotate and bend, and is capable of transferring axial compressive force ($N < 0$). The joint is not opened and exhibits linear elastic mechanical characteristics. The bending moment and elastic stiffness can be described as follows:

$$M = k_{\mathrm{el}} \Delta\phi \tag{4.29}$$

$$k_{\mathrm{el}} = Eb\frac{h^2}{12} \tag{4.30}$$

In the equation: $\Delta\phi$ is the rotational displacement of the joint; E is the Young's modulus of the concrete segment; h is the height of the joint, and b is the thickness of the joint, as illustrated in Fig. 4.17.

If the joint is eccentrically compressed, it may open up. When the magnitude of the bending moment $|M|$ is greater than $|N|\frac{h}{6}$, the mechanical behavior of the joint is described using a nonlinear relationship as follows:

$$M = 1/6\frac{(3|\Delta\phi|Ebh - 2\sqrt{2}\sqrt{|\Delta\phi|Ebh|N|})|N|\mathrm{sign}(\Delta\phi)}{|\Delta\phi|Eb} \tag{4.31}$$

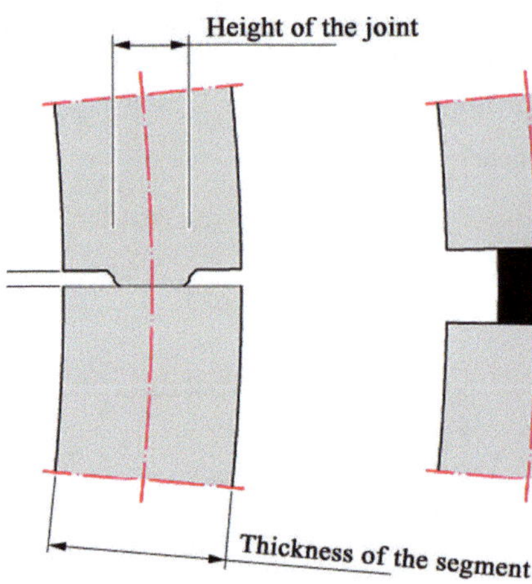

Fig. 4.16 Janssen equivalent beam joint model

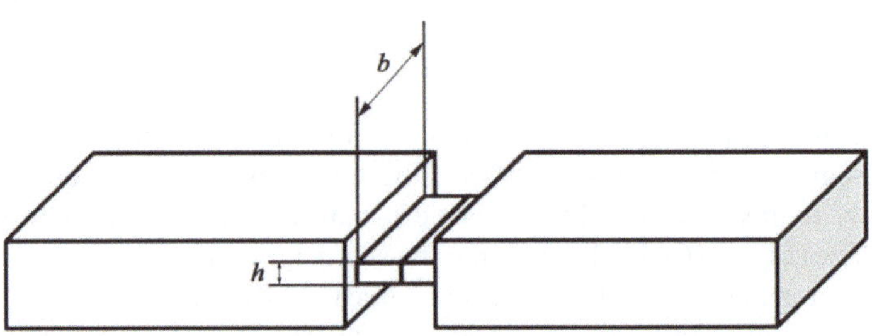

Fig. 4.17 Geometric parameters of the simplified Janssen equivalent beam joint model

$$k_t = 1/6\frac{N^2\sqrt{2}h}{|\Delta\phi|\sqrt{|\Delta\phi|Ebh|N|}} \tag{4.32}$$

If the joint is subjected to tension, then $N \geq 0$, $M = 0$.

Clearly, the stiffness of the joint k_{el} is only related to the modulus of elasticity E of the concrete and the height b and thickness h of the joint. Theoretically, the critical bending moment at which the joint exhibits both linear and nonlinear characteristics is $M = \frac{1}{6}Nh$.

4.3.2 Analysis and Calculation of Working Conditions

Excavating a foundation pit directly above a tunnel is an unloading behavior and is considered a high-risk engineering activity. Engineering cases demonstrate that the heave of the tunnel can be significant without control measures, which must be controlled to ensure the structural integrity and operational safety of tunnels. Currently, common auxiliary tunnel protection measures for foundation pit projects directly above tunnels include soil reinforcement, uplift piles, control of excavation size, and sectional excavation jumping. This section employs numerical simulation methods to analyze the effectiveness of commonly used auxiliary deformation control measures such as soil reinforcement in engineering, providing potential technical support for subsequent engineering design plans.

The calculation and analysis of working conditions consider combined effects of the foundation pit excavation size, soil reinforcement, and excavation time. Specific working conditions are shown in Table 4.16, and all unloading scenarios are a combination thereof.

4.3.3 Computational Parameters

The soil constitutive model selects the HS-Small model introduced in Sect. 4.2. The computational analysis software selects Zsoil, a finite element analysis software specialized for geotechnical engineering, which embeds a hardening soil small strain model. The tunnel lining segments are modeled using beam elements. The joints are simulated using an elastic nonlinear hinge element model embedded within the beam elements. The parameters for the hinge elements are determined based on comprehensive consideration of manual recommendations, research findings and computational experience. The calculation parameters for the soil model are determined based on the studies in Sect. 4.2. The calculation parameters for the soil layers are shown in Table 4.17. The parameters for the beam-hinge joint model are as presented in Table 4.18. The elastic modulus of the diaphragm walls and concrete supports is taken as 30.0 GPa, and the Poisson's ratio is assumed to be 0.2.

Table 4.16 Setting of working conditions for direct unloading

Excavation size		Unconfined compressive strength of reinforced soil q_u/MPa	Excavation time t/h	Uplift pile
Excavation depth h	Excavation width W			
$\frac{1}{3}H_0$	$\frac{1}{3}D$	Undisturbed soil (without soil reinforcement)	0	Available/None
$\frac{1}{2}H_0$	$\frac{1}{2}D$	0.8	8	
$\frac{2}{3}H_0$	D	1.0	12	
$\frac{3}{4}H_0$	$2D$	1.2	24	
H_0	$3D$	∞	∞	

Note 1. H_0 is the burial depth of the tunnel, with H_0 set at 15.0 m; D represents the outer diameter of the tunnel, with D equaling 6.2 m

2. The unconfined compressive strength of the reinforced soil $q_u = \infty$, meaning the strength of the reinforced soil reaches a large value, considered to be 2.0 MPa
3. The excavation time $t = 0$, indicating instantaneous unloading is complete. During the calculation and analysis, time is not considered. The setting of $t = 1$ merely indicates one analytical step. $t = \infty$ means the excavation time is considered in the analysis, and a large value is set for the time, such as $t = 30$d

Table 4.17 Parameters of the finite element calculation model

	c'/kPa	φ'/(°)	E_{oed}^{ref}/MPa	E_{50}^{ref}/MPa	E_{ur}^{ref}/MPa	E_0^{ref}/MPa	m	R_f	γ_{07}/(× 10^{-4})
①$_1$	10.0	30.0	3.41	3.81	23.81	75.2	0.65	0.9	3.2
②	4.0	30.0	3.41	3.81	23.81	75.2	0.65	0.9	3.2
③	2.0	31.4	2.61	2.85	18.52	56.2	0.65	0.9	3.2
④	4.0	24.0	1.92	2.02	13.94	41.2	0.65	0.9	3.2
⑤$_{11}$	7.0	25.1	2.82	3.10	21.32	106.7	0.65	0.9	3.2
⑤$_{12}$	6.0	30.0	3.33	3.71	30.26	151.3	0.80	0.9	3.2
⑤$_{31}$	6.0	32.2	4.78	5.73	34.41	172.0	0.80	0.9	3.2
⑤$_{32}$	5.0	32.4	4.34	5.21	31.23	156.2	0.80	0.9	3.2
⑤$_4$	22.0	31.7	6.10	7.32	43.93	219.7	0.80	0.9	3.2
⑦$_1$	4.7	29.0	7.03	8.85	43.67	220.0	0.50	0.9	3.2
⑦$_2$	5.0	29.0	7.03	8.85	43.67	220.0	0.50	0.9	3.2
⑨	5.0	35.0	7.03	8.85	43.67	220.0	0.50	0.9	3.2

Note In the calculations, p^{ref} is set to 100.0 kPa, and ν_{ur} is set to 0.2

Table 4.18 Parameters of the beam-joint model parameters

	E/GPa	ν	h/m	b/m
The beam-joint model	35.5	0.2	0.2	1.0

4.4 Analysis of Influencing Factors

4.4.1 *Excavation Size*

To investigate the relationship between the excavation size of the pit and tunnel deformation, the pit width is set to $\frac{1}{3}D$, $\frac{1}{2}D$, D, $2D$ and $3D$, and the excavation height is set to $\frac{1}{3}H_0$, $\frac{1}{2}H_0$, $\frac{2}{3}H_0$, $\frac{3}{4}H_0$ and H_0. The specific excavation sizes are shown in Table 4.19. The ratio of the excavation depth h to the overburden thickness H_0 above the tunnel is defined as the pit unloading ratio n, that is, $n = h/H_0$. The pit unloading ratio reflects the unloading amount relative to the tunnel burial depth. The size of the unloading directly affects the amount of heave deformation of the underlying tunnel.

Figure 4.18 illustrates the computational scenario for an excavation with dimensions W = 18.6 m and h = 7.5 m, located within the overburden zone above the tunnel defined by (45° + φ/2). In the diagram: s represents the amount of uplift of the tunnel, d represents the horizontal diameter of the tunnel after deformation, and l represents the vertical diameter of the tunnel after deformation. The variation in the horizontal diameter of the tunnel, defined as $\Delta d = d - D$, and the variation in the vertical diameter of the tunnel, defined as $\Delta l = l - D$. Δd is negative, indicating compression of the tunnel's horizontal diameter, while Δl is positive, indicating stretching of the tunnel's vertical diameter.

Table 4.19 Excavation sizes of the foundation pit

	Numerical value				
Excavation width W/m	2.1	3.1	6.2	12.4	18.6
Excavation depth h/m	5.0	7.5	10.0	11.2	15.0
Unloading ratio n	0.33	0.50	0.67	0.75	1.00

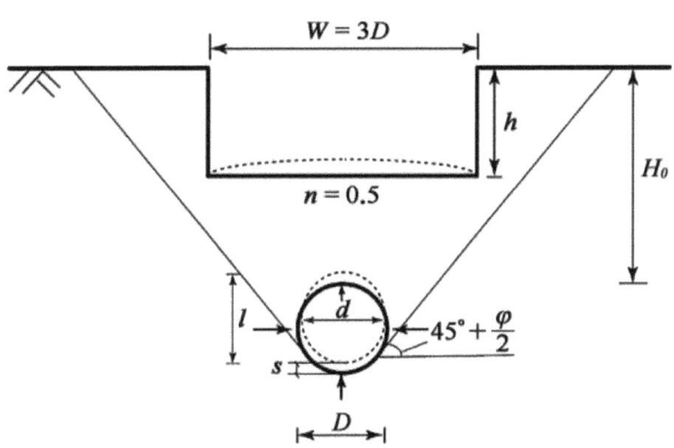

Fig. 4.18 Schematic diagram of the calculation condition for excavation of a pit above the tunnel ($W = 3D$, $h = ½H_0$)

When analyzing the impact of different excavation sizes of pits, soils inside the pit and around the tunnel are not reinforced, and no isolation protection measures are taken for the tunnel. The calculations use the parameters of the undisturbed soil. The results are shown in Figs. 4.19 and 4.20.

As can be seen from the figure, the tunnel heave deformation increases with growing excavation depth and width. When $W \leq D$, the amount of tunnel uplift is relatively small. However, as $W > D$, the tunnel uplift increases significantly with higher unloading ratio n. Using the parameters for undisturbed soil, when the excavation width is three times the outer diameter of the tunnel ($W = 3D$) and all the overburden soil above the tunnel is removed ($n = 1$), the tunnel uplift can reach up

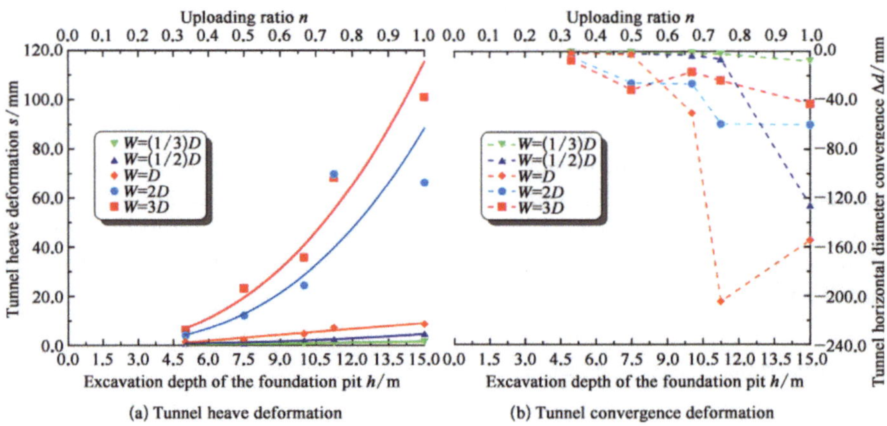

Fig. 4.19 Relationship between tunnel deformation and the size of the excavation pit

Fig. 4.20 Tunnel vertical heave deformation ($W \leq D$)

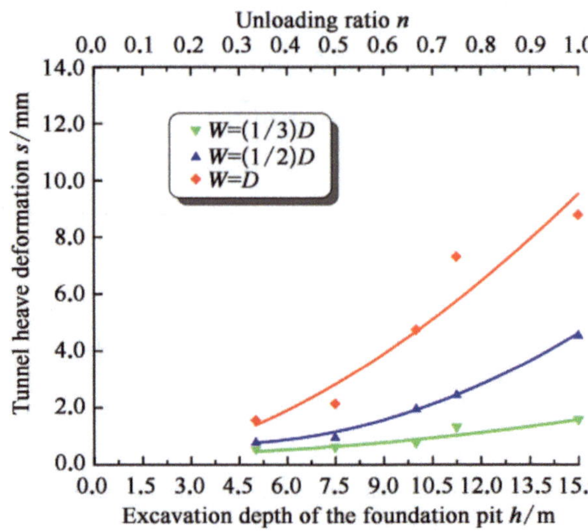

to 10.0 cm. Table 4.20 compares the rebound at the bottom of the pit and the uplift of the tunnel caused by the pit excavation when the tunnel burial depths are 15.0 and 10.0 m. After all the overburden soil is excavated, the uplift of the tunnel accounts for about 0.6–0.7% of the excavation depth of the pit.

The tunnel horizontal diameter convergence deformation pattern does not completely mirror the heave deformation, which implies it does not clearly exhibit a trend of increasing horizontal diameter convergence deformation with increasing unloading. For instance, the greatest convergence variation occurs when $W = D$ and $n = 0.75$. On the one hand, this is related to the computational conditions and the type of support structure configurations set up. On the other hand, it is associated with the tunnel lining segment joint model utilized in our study. The horizontal diameter convergence deformation depends on the stiffness of the tunnel segment joints. In this study, the tunnel joints are equated to concrete beams with weakened contact interfaces. Its correctness and rationality of this approach await further verification from actual engineering projects.

Moreover, based on the computational results of Fig. 4.19, when the excavation width of the pit equals the outer diameter of the tunnel ($W = D$), the tunnel uplift remains moderate. However, under the same unloading ratio conditions, the tunnel's convergence deformation is greater than those in the cases of $W = 2D$ and $W = 3D$. This is related to the setup of the enclosure structures in the computational analysis. When the width of the excavation pit above the tunnel is relatively small, not exceeding the outer diameter of the tunnel ($W \leq D$), the enclosure structures are commonly constructed using soil mixing walls (SMW construction method) or gravity dams, which are regarded as wall structures in numerical calculations and analyses. Since the reinforcement of the soil outside the tunnel is not yet considered, the displacement field of the soil caused by the excavation above can exacerbate the convergence deformation of the tunnel's horizontal diameter. For a single-circular assembled segment, the uplift at the bottom of the tunnel and the inward convergence of the horizontal diameter represent contradictory deformation trends, as shown in Fig. 4.21. Therefore, under such conditions, the tunnel deformation pattern is characterized by an uplift amount s that is far smaller than the transverse convergence deformation Δd.

Figure 4.22 illustrates the tunnel deformation pattern when the enclosure structure of the foundation pit is a diaphragm wall. Given that the width of the pit excavation is three diameters, the insertion ratio of the diaphragm wall is set to 1:1.5. With regard to for a tunnel buried 15.0 m deep ($H_0 = 15.0$ m) and with 75% of its loading

Table 4.20 The rebound at the bottom of the pit and the uplift of the tunnel (unit: mm)

	Tunnel available/none	$H_0 = 15.0$ m, $W = 3D$		$H_0 = 10.0$ m, $W = 3D$	
		The rebound at the bottom of the pit δ	The uplift of the tunnel s	The rebound at the bottom of the pit δ	The uplift of the tunnel s
$n = 1$	None	155.2	–	70.0	–
	Available	143.6	101.1	66.9	63.8

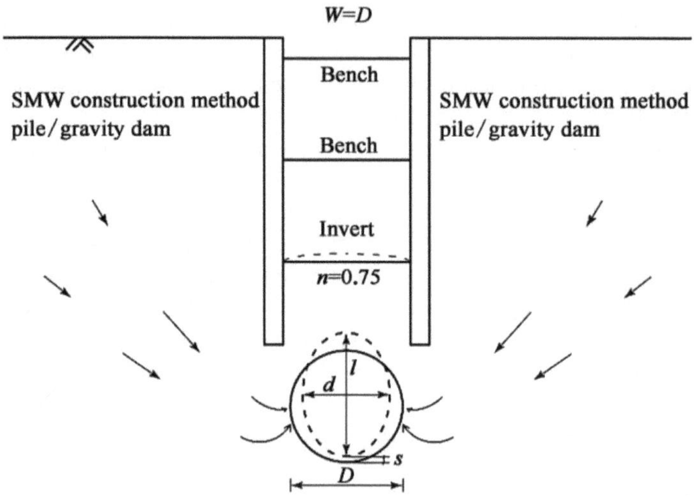

Fig. 4.21 Schematic diagram of tunnel deformation with upper foundation pit enclosure structure of SMW construction method pile/gravity dam ($W = D$)

removed ($n = 0.75$), the diaphragm wall extends below the tunnel's base, exceeding the depth at which the tunnel is buried. In this scenario, when the upper foundation pit is excavated, the diaphragm wall not only increases the soil's permeation path, but also acts as a barrier on both sides of the tunnel. It somewhat limits the development of the tunnel's horizontal deformation, resulting in the tunnel's uplift amount s being greater than the transverse convergence deformation Δd.

For the two scenarios mentioned, the unloading ratio n is the same, but the amount of unloading for the foundation pit with $W = 3D$ is much greater than that for $W = D$, hence $s|_{W=3D} \gg s|_{W=3D}$ can be obtained. Additionally, due to different enclosure structure configurations, $\Delta d|_{W=3D} < \Delta d|_{W=D}$ can be obtained. When comparing the

Fig. 4.22 Deformation of the tunnel with diaphragm wall of the upper foundation pit enclosure structure ($W = 3D$)

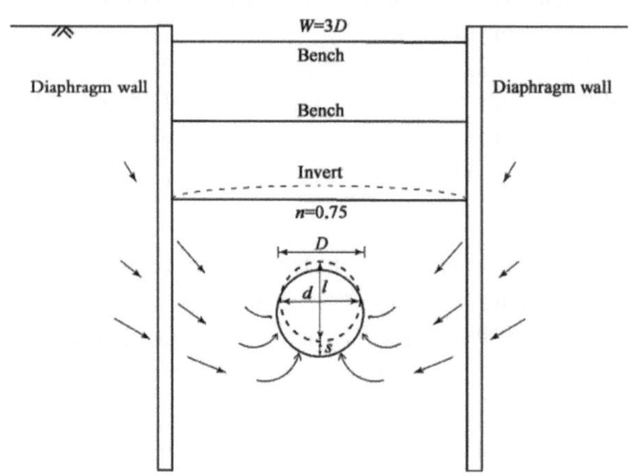

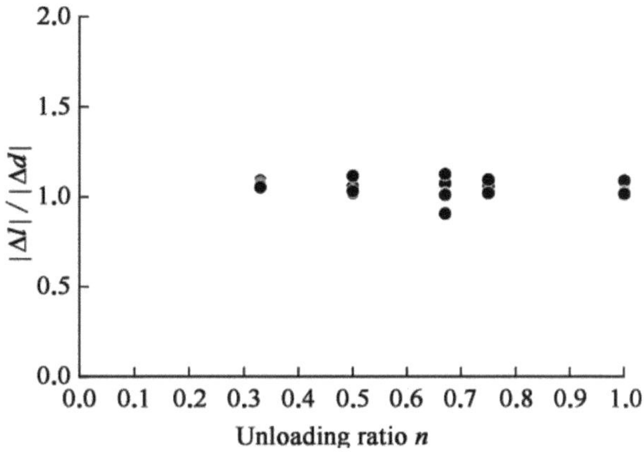

Fig. 4.23 The ratio of the convergence variation between the horizontal and vertical diameter of the tunnel for the upper foundation pit

vertical diameter variation of the tunnel in these two scenarios, $|\Delta l| \approx |\Delta d|$ can be found regardless of the pit width, suggesting that the tunnel's horizontal diameter compression is roughly equivalent to its vertical elongation.

Figure 4.23 illustrates the ratio of the convergence variation between the tunnel's horizontal diameter and its vertical diameter across all computational conditions. The results clearly involve $|\Delta l/\Delta d| \approx 1$.

If the excavation and unloading of the foundation pit directly above the tunnel are strictly zoned, segmented and symmetrical unloaded while keeping impacts within the structural tolerance of the tunnel (without causing strength failure of concrete segments or connecting bolts), the variation in the horizontal and vertical diameters of the tunnel segments will be internally balanced and harmonized. In short, the amount of compression in the horizontal diameter will be approximately equal to the amount of extension in the vertical diameter. Some studies have utilized the structural geometric features during the development of tunnel structure deformation to establish the relationship between the variation in the horizontal and vertical diameters of the shield tunnel as $\Delta l \approx -0.895\Delta d$ (Wang and Zhang 2013; Zhou et al. 2020), which closely matches this study's computational results.

Accordingly, tunnel deformation configuration affected by the upper unloading can be comprehensively examined based on the amount of uplift at the bottom s, transverse convergence deformation Δd, and vertical diameter variation Δl. After the tunnel assembly is completed, it gradually reaches a state of equilibrium within the soil. The top of the tunnel is subjected to the overburden pressure $\sigma_v = \gamma H_0$, which is relevant to the burial depth H_0. The waist of the tunnel is subjected to the lateral earth pressure $\sigma_h = K_0\sigma_v = (1 - \sin\varphi')\sigma_v$. In soft soil areas, if the lateral earth pressure coefficient $K_0 < 1$, the overburden stress σ_v at the top of the tunnel is greater than the horizontal earth pressure σ_h at the sides of the tunnel. Therefore, the initial convergence state of the tunnel after assembly should exhibit vertical compression and horizontal elongation, presenting a "lateral duck egg" behavior, as shown in Fig. 4.24.

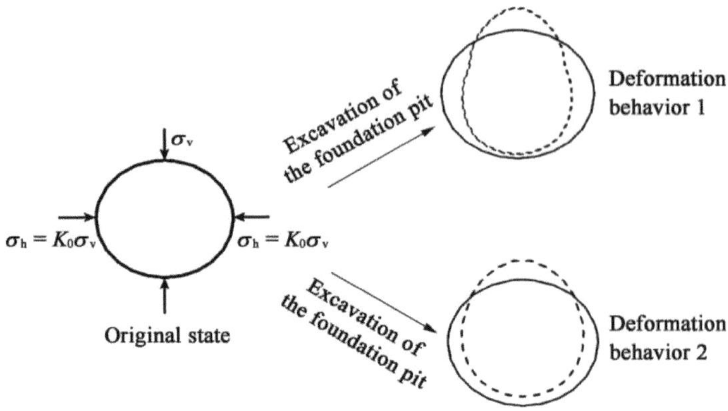

Fig. 4.24 Evolution of deformation configuration of subway tunnel caused by foundation pit excavation

If excavation is conducted above the tunnel, the overburden stress σ_V on the tunnel would decrease. The excavation and unloading lead to stress release in the soil layers traversed by the tunnel and underlying soil layers, causing soil rebound deformation. For one thing, the tunnel itself may undergo heave deformation due to the rebound of the soil. For another, it may accommodate the vertical heave deformation at the top and bottom of the tunnel by undergoing horizontal radial contraction, ultimately exhibiting a "vertical duck egg" deformation configuration characterized by vertical extension and horizontal compression, as illustrated in the deformation configuration schematic in Fig. 4.24.

The analysis of the two tunnel deformation configurations shown in Figs. 4.21 and 4.22 is summarized in Table 4.21, where the comparative results of the deformations are presented. Under the condition of $W = D$, the tunnel exhibits minor heave deformation but significant transverse convergence deformation, resulting in a tunnel shape as shown in deformation configuration 1 in Fig. 4.24. Under the condition of $W = 3D$, the tunnel produces greater heave deformation and substantial lateral convergence, though less than the uplift, as depicted in deformation configuration 2 in Fig. 4.24. In terms of the ellipticity of the tunnel, the ellipticity of deformation configuration 2 is evidently better. The deformation caused by the upper excavation is relatively uniform, which is beneficial for the structural safety and stability of the tunnel. This is extremely important for subway tunnels in service. In contrast, the tunnel with deformation configuration 1 suffers from severe local lateral compression, making it prone to water leakage through the longitudinal joints. This is highly detrimental to the structural safety. The subsequent cost of operational maintenance and management would also be higher. Therefore, it's essential to control and avoid this type of deformation as much as possible.

Considering $\delta = |\Delta d|/|s|$, the magnitude of the ratio δ can reflect the uniformity of the tunnel deformation. The vertical deformation s at the bottom of the tunnel is a positive value indicating tunnel uplift, and a negative one indicates tunnel settlement.

Table 4.21 Comparison of tunnel deformation results

	$n = 0.75$	
$W = D$	$s\vert_{W=D} = \Delta d\vert_{W=D}, \Delta d\vert_{W=D} \approx \Delta l\vert_{W=D}$	$s\vert_{W=3D} \gg s\vert_{W=D}$
$W = 3D$	$s\vert_{W=3D} > \Delta d\vert_{W=3D}, \Delta d\vert_{W=3D} \approx \Delta l\vert_{W=3D}$	$\Delta d\vert_{W=3D} < \Delta d\vert_{W=D}$

If the ratio δ is a large number, it means that the transverse convergence deformation of the tunnel is much larger relative to the tunnel uplift, indicating that the tunnel experiences severe horizontal stretching or compression. This deformation behavior is generally unfavorable. When $\delta \to 1$, it implies that $|\Delta d| \approx |s|$, indicating the tunnel deformation can be directly assessed based on the magnitude of the deformation values. When $\delta < 1$ or $\delta \to 0$, it suggests that the tunnel's cross-sectional convergence deformation is relatively small compared to the tunnel's uplift or settlement, with vertical deformation being the major form of deformation. In such cases, a larger value of s indicates more significant uplift or settlement of the tunnel. This deformation requires an examination of the uplift and settlement deformations along the longitudinal direction of the tunnel and the differential uplift and settlement deformations between the ring segments.

Figure 4.25 shows the layout of monitoring points on the segments of operational subway tunnels. Settlement monitoring points are typically placed on the track bed, and horizontal radial convergence monitoring points are positioned around the waist of the tunnel's horizontal diameter. After casting, the track bed integrates with the tunnel. If, during operation, the track bed and the tunnel remain intact, the variation in uplift and settlement from the track bed monitoring points can be considered as indicative of the tunnel's uplift and settlement deformations. On the basis of the aforementioned tunnel deformation analysis, the deformation configuration of the tunnel segments can be evaluated using the monitoring data of tunnel settlement and convergence deformation. Combined with an on-site investigation of defects, this allows for an analysis of the service condition of the tunnel.

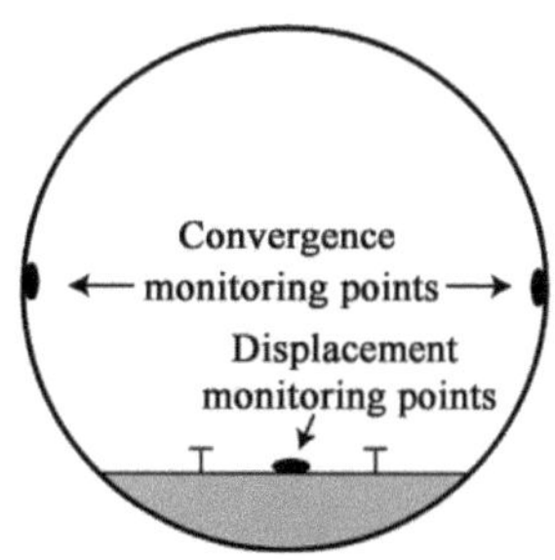

Fig. 4.25 The schematic diagram of the layout of monitoring points on the segments of operational subway tunnels

4.4.2 Soil Reinforcement

According to general design requirements for shield tunnels, once assembled into rings under external impacts, the tunnel is permitted to have its longitudinal and circumferential joints open up to 6 mm without leaking, and the cumulative change in diameter should be less than 0.5% of the tunnel's diameter. Therefore, for a single-circular tunnel with an external diameter D of 6.2 m, post-assembly deformations should be controlled within 3.1 cm. The deformation protection requirements for the normal operation of subway shield tunnels are very strict. According to the standard "Technical Specifications for Structural Safety Protection of Urban Rail Transit" (CJJ/T 202—2013), the vertical displacement and radial convergence of the tunnel should not exceed 20 mm. The calculations from the preceding section indicate that if a foundation pit is excavated above the tunnel without soil reinforcement or other controlling measures, the tunnel deformation will surpass its allowable limit, posing a challenging controlling situation that could significantly jeopardize the operational safety of the structure.

(1) Portal reinforcement of the soil outside the tunnel

The soil around the tunnel can be reinforced by using construction techniques such as grouting, triaxial mixing piles, rotary jet piles, or high-frequency vibrating casting piles. The reinforcement range is shown as the shaded area in Fig. 4.26. The upper reinforced soil has a thickness of 3.0 m and is located 0.5 m from the top of the tunnel. On the sides, the reinforced soil is 4.0 m wide and is located 1.5 m from the tunnel. The lateral reinforcement extends 3.0 m downward from the bottom of the tunnel, forming a "portal" shaped reinforcement around the tunnel.

To reinforce the soil, cement content should reach approximately 15%. The reinforced soil has an unconfined compressive strength q_u of at least 0.8 MPa after 28 days. The calculation parameters for the soil within the reinforcement range are

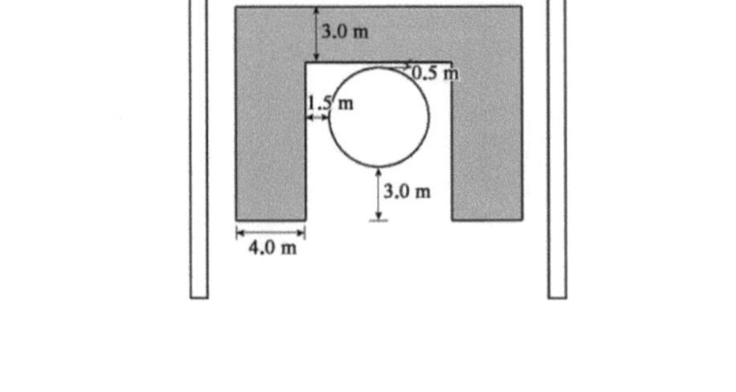

Fig. 4.26 Portal reinforcement of the soil outside the tunnel

presented in Table 4.22. The calculation results of tunnel deformation under different excavation sizes are listed in Fig. 4.27.

The computational results from Fig. 4.27 show that, for the soil outside the tunnel undergoing portal reinforcement, the deformation of the tunnel caused by the excavation of the upper pit mostly manifests as the vertical uplift *s* being greater than the transverse convergence deformation Δd. This is opposite to the tunnel deformation behavior when the soil is not reinforced. This indicates that portal reinforcement around the soil has influenced the deformation behavior of the tunnel induced by the upper pit excavation. Figure 4.28 compares the ratio δ of the transverse convergence deformation Δd and the uplift *s* before and after reinforcement. It can be clearly seen from the figure that after the tunnel undergoes portal reinforcement, δ changes from greater than 1 to less than 1. The portal reinforcement effectively suppresses the transverse convergence deformation caused by the upper unloading, and the tunnel mainly exhibits vertical uplift.

Tunnel heave deformation can be further controlled by limiting the rebound deformation of the soil due to pit excavation. The horizontal convergence and vertical deformation of the tunnel segments do not occur and develop independently. It is necessary to consider deformation control measures comprehensively to prevent unfavorable deformation modes.

Table 4.22 Parameters of the reinforced soil model

	c'/kPa	φ'/(°)	E_{oed}^{ref}/ MPa	E_{50}^{ref}/ MPa	E_{ur}^{ref}/ MPa	E_{0}^{ref}/ MPa	m	R_f	$\gamma_{0.7}$/(× 10^{-3})
Reinforced soil	60	25	15.00	15.00	80.00	170.0	0.12	0.9	2.0

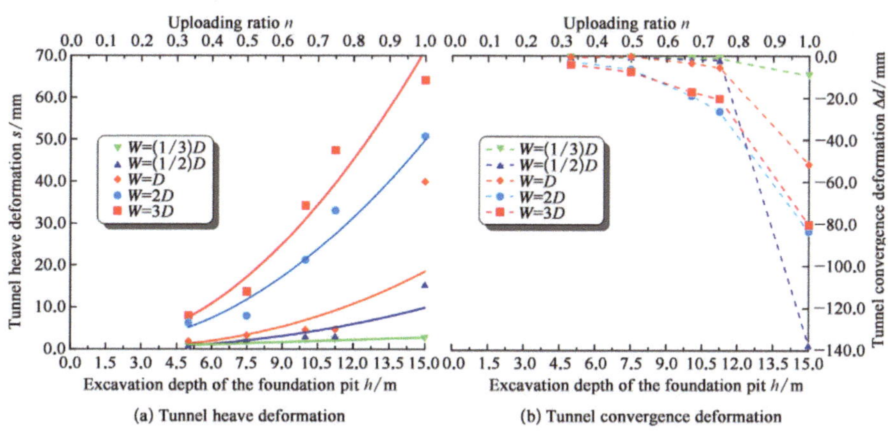

Fig. 4.27 The excavation of the foundation pit with portal reinforcement impacts tunnel deformation

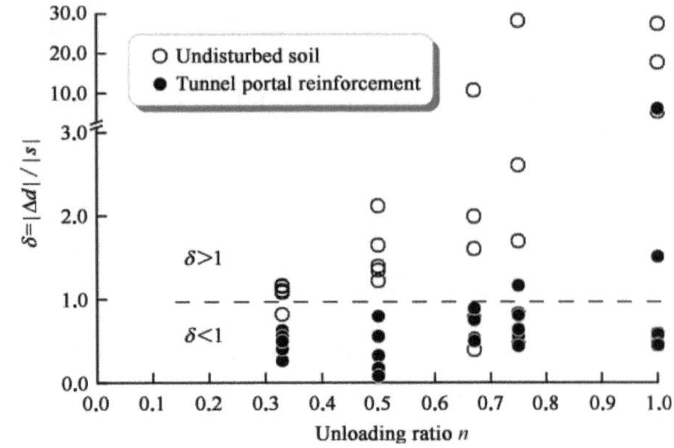

Fig. 4.28 The ratio δ of the convergence deformation Δd and the heave deformation s before and after reinforcement

(2) Full-scale reinforcement of the pit above the tunnel

Indeed, portal reinforcement of the soil surrounding the tunnel can control, to a certain extent, the deformation impacts of pit excavation on the tunnel. However, when the excavation width W exceeds the outer diameter D and the unloading ratio n is high, the deformation of the tunnel remains significant, exceeding 20 mm, and thus fail to meet the subway's operational and maintenance standards. Now, the full-scale reinforcement within the pit above the tunnel is conducted using techniques such as grouting and mixing piles, as illustrated in Fig. 4.29. The calculation results can be seen in Fig. 4.30, Tables 4.23, and 4.24.

As shown in Fig. 4.30, the uplift and convergence deformations of the tunnel are further effectively controlled after full-scale reinforcement of the soil within the pit, where the uplift s exceeds the transverse convergence deformation Δd. Both s and Δd increase with the unloading ratio n. The ratio of the uplift s to the excavation depth h varies between 1.00 and 2.72‰. The variations in the horizontal and vertical

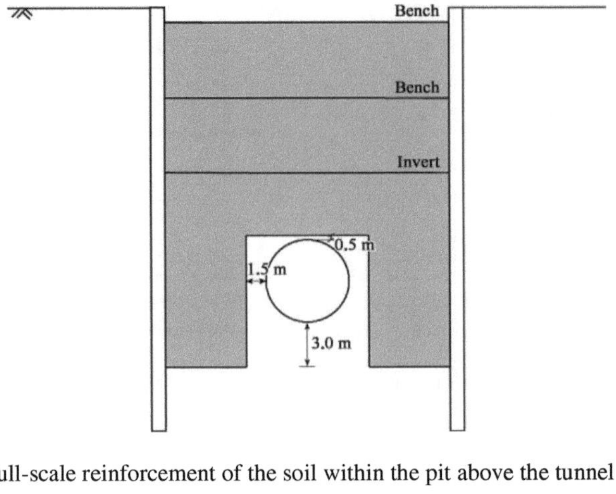

Fig. 4.29 The full-scale reinforcement of the soil within the pit above the tunnel

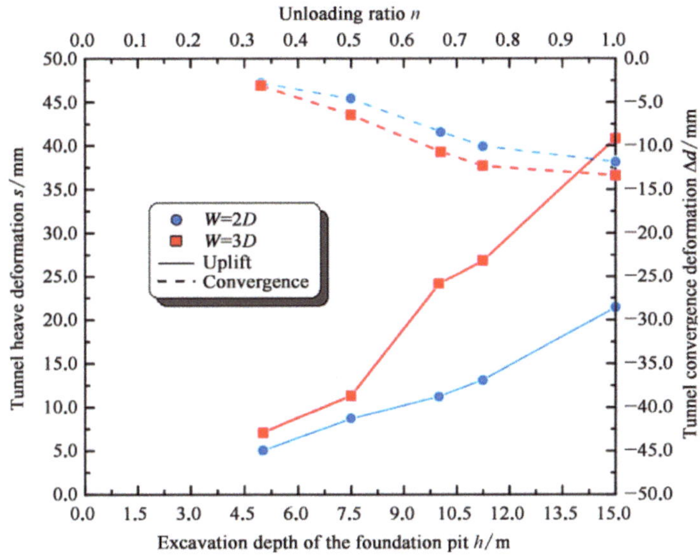

Fig. 4.30 The excavation of the pit with full-scale reinforcement impacts tunnel deformation

Table 4.23 The tunnel deformation under the condition of full-scale reinforcement and $W = 2D$

	Numerical value				
Unloading ratio n	0.33	0.50	0.67	0.75	1.00
Vertical uplift s/mm	5.0	8.7	11.3	13.2	21.4
Horizontal diameter convergence Δd/mm	−2.8	−4.6	−8.4	−10.1	−11.9
Vertical diameter convergence Δl/mm	3.1	5.0	8.7	10.7	12.4

Table 4.24 The tunnel deformation under the condition of full-scale reinforcement and $W = 3D$

	Numerical value				
Unloading ratio n	0.33	0.50	0.67	0.75	1.00
Vertical uplift s/mm	7.1	11.4	24.2	26.9	40.8
Horizontal diameter convergence Δd/mm	−3.1	−6.5	−10.7	−12.3	−13.4
Vertical diameter convergence Δl/mm	3.4	6.9	11.2	12.9	14.0

diameters of the tunnel, as shown in Fig. 4.31, are consistent with the aforementioned deformation pattern, complying with $|\Delta l| \approx |\Delta d|$.

Figure 4.32 compares the impact of pit excavation on tunnel deformation under two conditions: full-scale reinforcement of the soil within the pit and portal reinforcement of the soil outside the tunnel only. After the full hall reinforcement of the soil within the pit, when the unloading ratio $n < 1$, the average heave deformation of the tunnel is reduced by about 35%, and the convergence deformation by about 40%.

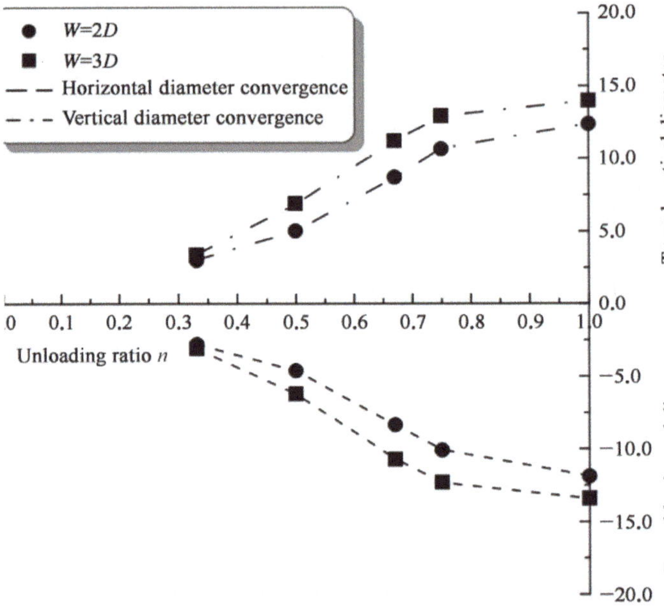

Fig. 4.31 Convergence variations in the horizontal and vertical diameters of the tunnel after full-scale reinforcement of the soil within the upper pit

The deformation control effect is even more significant when the unloading ratio $n = 1$.

(3) Strength of soil reinforcement

Typically, the design of pit enclosure structures has certain requirements for soil reinforcement, including construction technique for ground reinforcement, cement

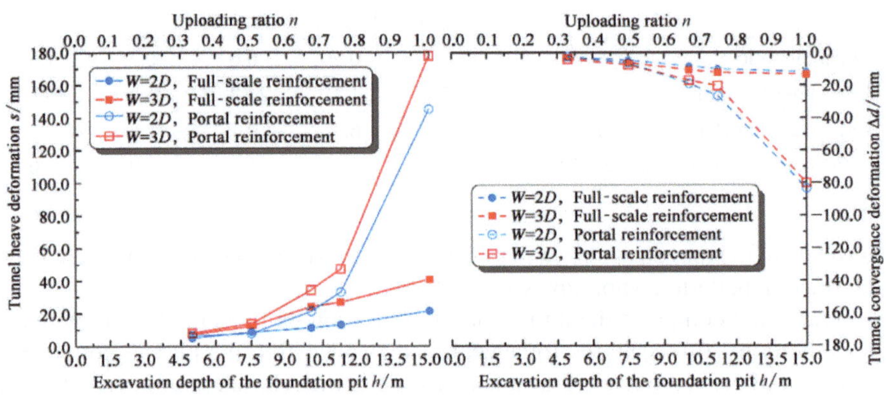

(a) Comparison of tunnel heave deformation (b) Comparison of tunnel convergence deformation

Fig. 4.32 Impact of pit excavation on tunnel deformation under soil reinforcement conditions

Table 4.25 The requirements and calculation parameters for soil reinforcement

	Cement content (%)	Unconfined compressive strength q_u/MPa	Unloading modules E_{ur}^{ref}/MPa	Permeability coefficient k/(cm·s^{-1})	Construction technique
Cement-reinforced soil	15	≥0.8	80	≤ 10^{-7}	Grouting, triaxial, mixing piles
	20	≥1.0	100		
	25	≥1.2	120		
	40–50	≥2.0	200		

Table 4.26 Comparison of the tunnel deformation results under the condition of $W = 3D$ and $n = 0.5$

	Numerical value				
Unconfined compressive strength q_u/MPa	Original soil	0.8	1.0	1.2	2.0
Vertical uplift s/mm	23.2	11.4	11.2	10.9	10.8
Horizontal diameter convergence Δd/mm	−31.2	−6.5	−5.8	−4.9	−3.2
Vertical diameter convergence Δl/mm	32.2	6.9	6.2	5.3	3.5

content, unconfined compressive strength of the reinforced soil, cement grade, etc. This section considers several different strength parameters for reinforced soil, as shown in Table 4.25. The calculations take into account both full-scale reinforcement of the soil within the pit and portal reinforcement of the soil outside the tunnel for conditions where the unloading ratio $n = 0.5$ and the excavation width $W = 3D$, with results presented in Table 4.26.

Based on the results from Table 4.26, compared to the unreinforced soil (i.e., the original soil), after soil reinforcement ($q_u \geq 0.8$ MPa), the uplift s caused by the excavation of the pit above the tunnel is reduced by up to 50%, and the transverse convergence deformation Δd also decreases by more than 50%. As illustrated in Fig. 4.29, it is evident that full-scale reinforcement of the soil within the pit and portal reinforcement of the soil on both sides of the tunnel, combined with meeting the strength requirements of soil reinforcement, can effectively control the tunnel uplift and convergence deformations induced by the upper pit excavation.

4.4.3 Uplift Pile

To better control the adverse deformations of a tunnel caused by the excavation of a pit directly above it, it is considered to install uplift piles on both sides of the tunnel, forming a portal frame with the pit's bottom slab, as illustrated in Fig. 4.33. The uplift piles are designed as drilled cast-in-place piles with a diameter of 850 mm and a center-to-center spacing of 5000 mm. The top of the piles is embedded into the bottom slab by about 0.1 m, and the length of the piles is controlled to reach the bearing stratum at layer ⑦$_2$, with an embedding depth of 5.0–7.0 m. They are modeled using beam elements with linear elasticity, with an elastic modulus of 30.0 GPa and a Poisson's ratio of 0.2. Computational results can be seen in Fig. 4.34, Tables 4.27, and 4.28.

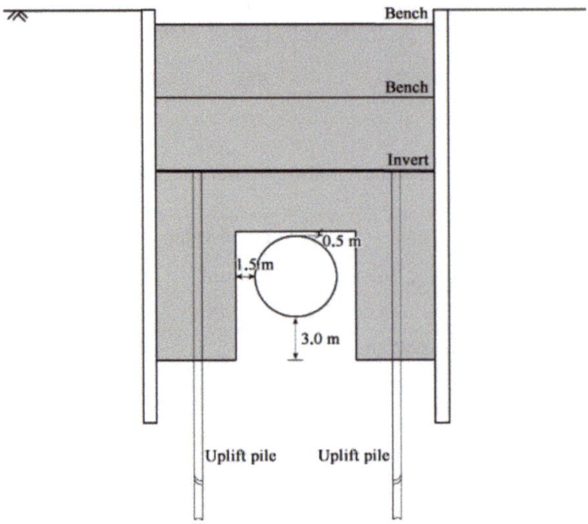

Fig. 4.33 Portal reinforcement of the tunnel uplift pile

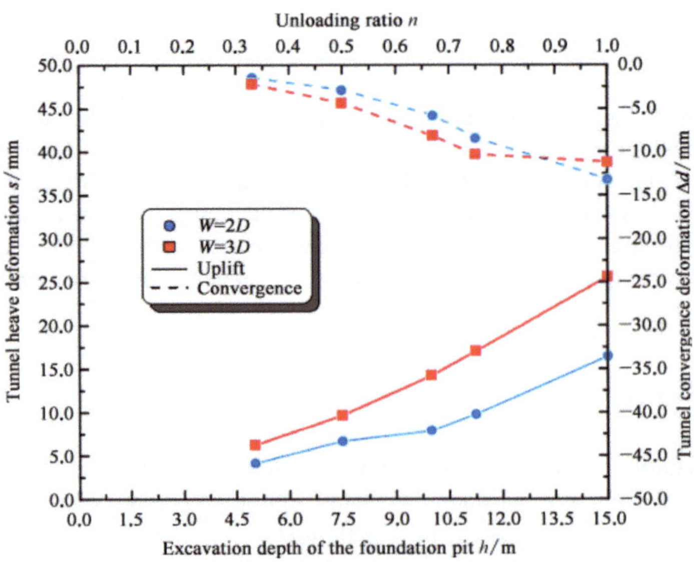

Fig. 4.34 Impact of pit excavation on tunnel deformation with uplift piles set up outside the tunnel

Table 4.27 The tunnel deformation under full-scale reinforcement and uplift piles for the condition of $W = 2D$

	Numerical value				
Unloading ratio n	0.33	0.50	0.67	0.75	1.00
Vertical uplift s/mm	4.1	6.7	7.9	9.8	16.4
Horizontal diameter convergence Δd/mm	−1.4	−2.9	−5.8	−8.4	−13.2
Vertical diameter convergence Δl/mm	1.6	3.2	6.3	8.9	13.7

Table 4.28 The tunnel deformation under full-scale reinforcement and uplift piles for the condition of $W = 3D$

	Numerical value				
Unloading ratio n	0.33	0.50	0.67	0.75	1.00
Vertical uplift s/mm	6.2	9.7	14.2	17.0	25.6
Horizontal diameter convergence Δd/mm	−2.2	−4.3	−8.2	−10.3	−11.2
Vertical diameter convergence Δl/mm	2.4	4.7	8.7	11.0	11.9

The comparison results of tunnel deformation under different pit excavation sizes with and without uplift piles are shown in Fig. 4.35. It can be seen from the figure that uplift piles form a portal protective hoop with the pit's bottom slab, which largely enhances tunnel protection. They not only reduce the tunnel's heave deformation but also limit its transverse convergence deformation. The installation of uplift piles results in approximately a 25% reduction in both uplift and convergence deformations of the tunnel.

Figure 4.36 displays the deformation configuration of the tunnel with uplift piles installed. It is clearly visible from the figure that under the influence of the excavation

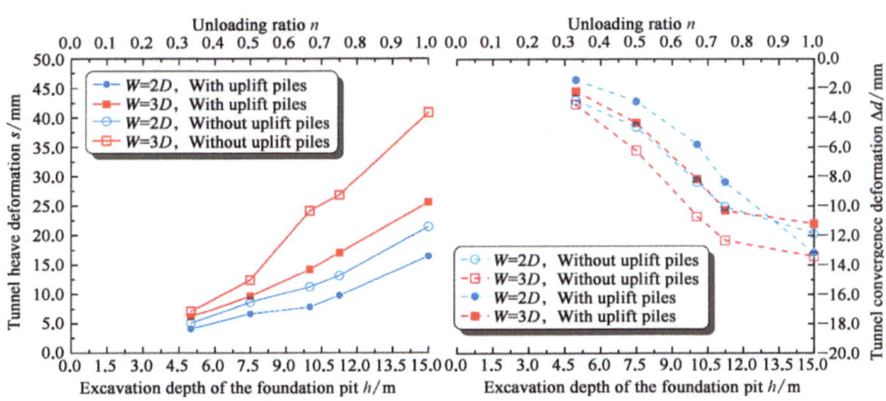

Fig. 4.35 Comparison of tunnel deformation from pit excavation with and without tension piles

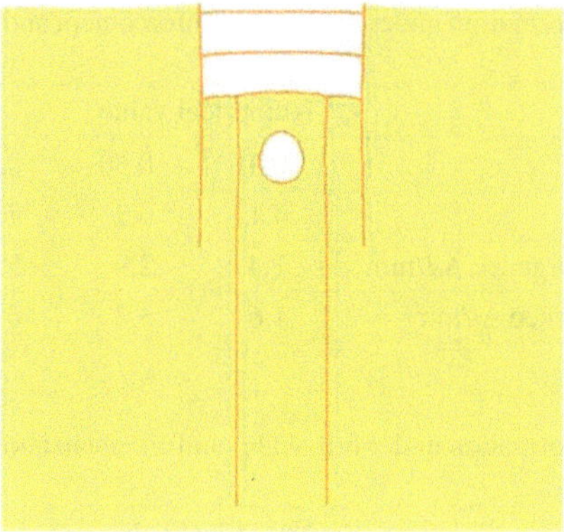

Fig. 4.36 Deformation configuration of the tunnel under the influence of upper pit excavation

of the pit above, the deformation trend of the underlying subway tunnel progresses towards a "vertical duck egg" shape.

Figure 4.37 presents the soil displacement deformation pattern under the excavation sizes of $W = 3D$, and $n = 0.5$. It can be observed that the soil behind the underground wall is a settlement zone, while the soil beneath the pit bottom is an uplift zone. The area approximately twice the depth of the pit excavation behind the underground wall is the main settlement influence zone. Based on the tunnel depth and the soil stratification, the primary overburden influence zone above the tunnel can be approximated as the region defined by an angle of approximately twice the $(45^\circ + \frac{\varphi}{2})$ from the tunnel crown . Accordingly, it is essential to carry out targeted maintenance and monitoring during the operational phase of the subway tunnel.

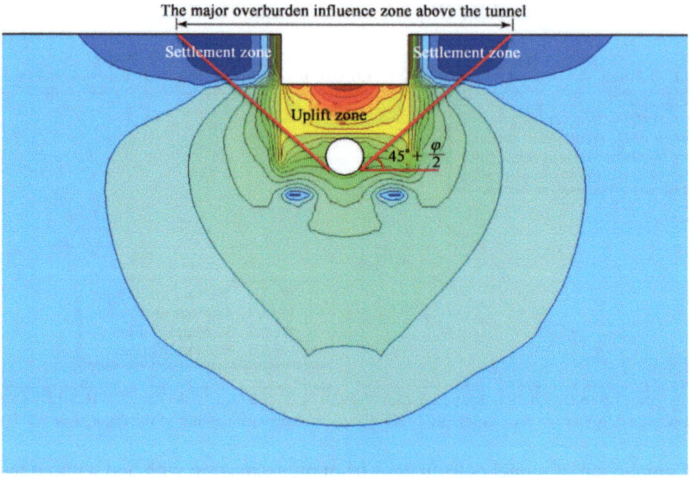

Fig. 4.37 The major overburden influence zone above the tunnel

According to the aforementioned results (Fig. 4.34), when conducting pit excavation directly above the tunnel, the following reinforcement and controlling measures must be implemented: portal reinforcement of the soil outside the tunnel, full-scale reinforcement within the pit, and installation of uplift piles on both sides of the tunnel. These measures can control the ratio of the tunnel uplift s to the pit excavation depth h within the range of 0.78–1.71‰, with an average value of 1.16‰. Previous studies have statistically analyzed the measured data of tunnel heave deformation caused by pit excavation. The studies concluded that within an unloading ratio n range of 0.40–0.90, the maximum tunnel uplift s_{max} to pit excavation depth h ratio varies from 0.63 to 3.00‰ in soft soil areas.

4.4.4 Excavation Time

Shanghai soft soil is characterized by its nonlinear rheological properties. Engineering practice has proven that the rebound amount of the soil layers at the bottom of the foundation pit is related to time when excavating foundation pits in soft soil areas. Slower excavation progress and longer exposure time after excavation lead to greater rebound in the pit bottom soil. If a foundation pit lies directly above a tunnel, excessive soil rebound will inevitably result in an overlarge heave deformation of the tunnel, subsequently triggering structural safety diseases. Now considering the time factor, this study analyzes the impact of different excavation durations on tunnel deformation.

For ease of discussion regarding the influence of time, the analysis is conducted for a pit size of $W = 3D$ and $n = 0.5$. It is assumed that the first strut of the pit is set at 1.0 m below the ground surface, and the excavation goes directly to the bottom in the second soil layer at 6.5 m, with an excavation volume of about 120 m^3. Five excavation durations are set: 0 ($t = 1$ represents 1 analysis step), 8, 12, 24 h, and ∞ ($t = 30$d), to compare and analyze the deformation of the tunnel after the completion of excavation. The computational analysis involves full-scale reinforcement of the soil within the pit, portal reinforcement of the tunnel, and the setup of uplift piles. In Fig. 4.38, an excavation time of $t = 1$ is considered as disregarding the time factor. Different excavation durations stand for various soil unloading rates. The longer the time, the slower the excavation rate. Conversely, shorter times are equivalent to instantaneous unloading. All five excavation durations are assumed to involve uniform unloading.

Figure 4.39 compares the deformation of the tunnel under the condition of $W = 3D$, $n = 0.5$, taking into account the impact of the excavation time factor, where the horizontal axis is on a logarithmic scale. From the figure, it is evident that as the excavation rate decreases and the excavation time increases, both the uplift and convergence deformations of the tunnel develop significantly. Specifically, the influence on the tunnel's heave deformation exceeds that of convergence deformation. The longer the excavation time, the greater the uplift of the tunnel. For a tunnel where excavation is completed in 24 h, the uplift rises by about 69%, accounting for 2.17‰ of the excavation depth. For a tunnel where excavation is completed in 30 days, the

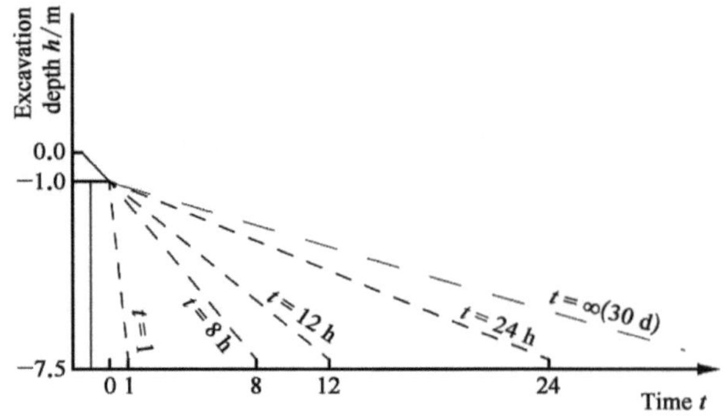

Fig. 4.38 Unloading and excavation time

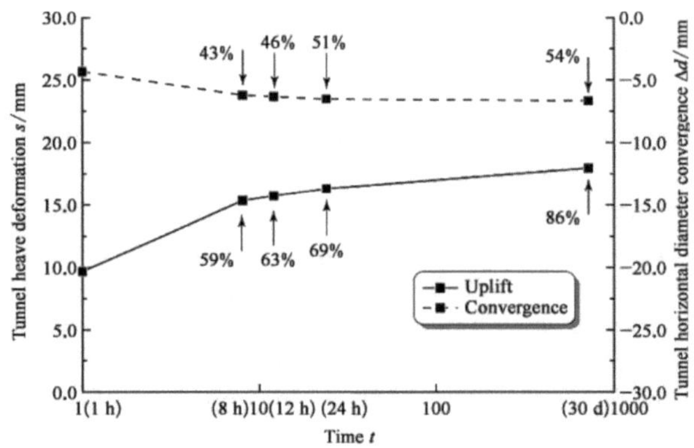

Fig. 4.39 Tunnel deformation at different excavation times

uplift increases by about 86%, amounting to approximately 2.4‰ of the excavation depth.

Figure 4.40 illustrates the duration curves of tunnel heave deformation for different excavation durations. The figure demonstrates that starting from the excavation of the second soil layer, as the soil continues to be unloaded, the tunnel heave deformation progressively increases. A faster excavation rate results in a relatively smaller final uplift amount. After the excavation is completed and the bottom slab is poured, the development of heave deformation is effectively restrained. Therefore, for pit projects located directly above a tunnel, it is advisable to minimize the soil excavation time as much as possible and to complete the bottom slab pouring as quickly as possible. These measures can effectively limit the development of tunnel heave deformation.

Considering the extreme excavation time $t \to \infty$, it can be observed from Fig. 4.39 that as the excavation time is extended, the uplift of the tunnel still grows slowly (with the excavation time set at 30 days, the tunnel heave deformation is approximately 18.0 mm). However, compared to a condition where the excavation is completed in a shorter span of time, the uplift increment of the tunnel is considerably larger. This suggests that completing the excavation process within a shorter duration is advantageous in controlling tunnel deformation. Consequently, when excavating foundation pits directly above the tunnel, it is crucial to consider the adverse effects of time

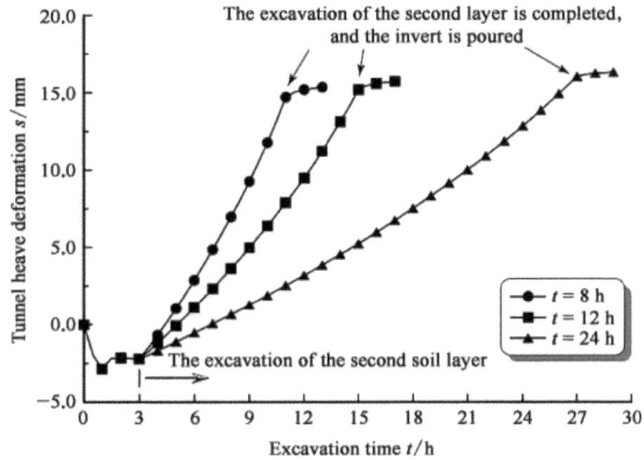

Fig. 4.40 The duration curve of tunnel heave deformation under Influence of different excavation times

factors on soil rebound amount. Measures such as accelerating the pace of excavation and support, reducing the exposure time of the pit bottom, and promptly casting the cushion layers, inverts, and other subterranean structures should be adhered to.

4.5 Conclusion

This chapter expounds upon the HSS constitutive model for typical soil layers in Shanghai and its parameter determination methods. The main focus of the analysis is the heave deformation of the tunnel and its horizontal and vertical convergence deformations caused by the excavation and unloading of the pit directly above tunnels. Through numerical analysis, the main factors affecting tunnel deformation are examined, including the plan sizes of the pit excavation, the form and scope of soil reinforcement, the setting of uplift piles, the excavation duration, etc. By employing a developed finite element model and through comparative analysis, the following preliminary conclusions have been drawn:

① The multi-hinge joint model adopted for the tunnel segment structure can effectively simulate the deformation configuration of the tunnel under the upper unloading. Affected by the excavation and unloading of the pit directly above tunnels, the tunnel's deformation exhibits a vertical ovalization deformation characterized by vertical stretching and horizontal compression. Furthermore, the tunnel segment structure, within its strength capacity, demonstrates balanced and coordinated variations in its horizontal and vertical diameters.

② After soil reinforcement, the deformation configuration of the tunnel under the effect of unloading is changed, primarily manifesting as vertical heave deformation. For the tunnel lining segments themselves, uplift (or settlement) and convergence deformations do not occur and develop independently. The trends in diameter variations in both directions are contradictory, serving as an adaptation to the deformations produced in the soil due to stress release. Specifically, for the upper unloading, the main mode of deformation induced in the tunnel lining segments is heave deformation. Due to the assembled structure, the ellipticity of the segments change, accompanied by transverse convergence deformation. This suggests that the deformation of the tunnel lining coincides with the overall heave deformation following soil rebound and changes in its own diameter.

③ As the unloading ratio n of the soil mass above the tunnel increases, the tunnel's heave deformation shows an upward trend. When the excavation width W of the foundation pit is less than the tunnel's outer diameter D, the growth in deformation is limited. However, when $W > D$ and $n > 0.5$, the heave deformation shows a substantial increase, necessitating control measures to restrict tunnel deformation. Portal reinforcement of the soil outside the tunnel can suppress the development of the tunnel's transverse convergence deformation and improve its deformation configuration. After full-scale reinforcement of the soil within the pit, the tunnel's heave deformation is reduced by about 35% on average, and the convergence deformation is reduced by about 40%.

④ Full-scale reinforcement of the soil within the pit above the tunnel can effectively limit the soil rebound, thereby further controlling tunnel deformation. Compared to the unreinforced soil, i.e., the original soil, after soil reinforcement ($q_u \geq 0.8$ MPa), the heave deformation s of the tunnel caused by the excavation of the pit above is reduced by up to 50%, and the transverse convergence deformation Δd is also reduced by over 50%. If the strength requirements for the soil reinforcement are strictly met, the effect on controlling both the uplift and convergence deformations of the tunnel induced by the excavation of the pit above would be remarkable.

⑤ Installing tension piles on both sides of the tunnel to form a portal protective hoop with the pit's bottom slab significantly enhances the tunnel's protection, further reducing both the uplift and convergence deformations of the tunnel by approximately 25%. For excavating a pit directly above the tunnel, take into account the following measures, including portal reinforcement of the soil outside the tunnel, full-scale reinforcement of the soil within the pit, and installing uplift piles on both sides of the tunnel. In consequence, the ratio of the tunnel uplift s to the excavation depth h can be controlled within a range of 0.78–1.71‰, with an average value of 1.16‰.

⑥ Controlling the excavation time of a pit is beneficial for controlling the development of tunnel deformation. The longer the excavation time, the greater the tunnel deformation will be. The impact of time factors on tunnel heave deformation is more substantial than on transverse convergence deformation. The convergence trend of tunnel deformation over time relates fundamentally to the consolidation process following soil disturbance.

⑦ When a foundation pit is excavated above the tunnel, deformations show strong correlations with multiple factors such as the foundation pit size, the form of the pit retaining system, the range and solidity of soil reinforcement, and excavation time. Deformation control measures demonstrate interdependent effects rather than isolated functions, requiring comprehensive consideration based on specific project conditions during implementation.

References

Bao HD, Wu JB, Xu CS (1989) Determination of dynamic parameters of cohesive soil in Shanghai area by resonant column test. Site Investig Sci Technol 02:18–23

Benz T (2006) Small strain stiffness of soils and its numerical consequences. University of Stuttgart, Stuttgart

Chen F (2011) Study on adaptability of constitutive models for typical strata in Wuxi metro foundation pits. Master's thesis, Tongji University, Shanghai

Drucker DC, Prager W (1952) Soil mechanics and plastic analysis or limit design. Q Appl Math 10(02):157–165

Drucker DC, Gibson RE, Henkel DJ (1957) Soil mechanics and work-hardening theories of plasticity. Proc ASCE Tran 122:338–346

Duncan JM, Chang CY (1970) Nonlinear analysis of stress and strain in soils. J Soil Mech Found Div ASCE 96(05):1629–1653

Fan YY, Guo XH, Deng ZJ et al (2014) Impact analysis of overlying loading/unloading on shield tunnel deformation. Constr Technol 43(07):107–109

Hardin BO, Drnevich VP (1972) Shear modulus and damping in soils. J Soil Mech Found Div 98(07):667–692

Janssen P (1983) Tragverhalten von Tunnelausbauten mit Gelenktuebbings. University of Braunschweig, Department of Civil Engineering, Institute for Structural Analysis

Jiang J (2009) Experimental study on small-strain characteristics and long-term deformation behavior of Shanghai soft clay. PhD dissertation, Tongji University, Shanghai

Kondner RL (1963) Hyperbolic stress-strain response: cohesive soils. J Soil Mech Found Div ASCE 89(01):115–143

Liang FY, Jia YJ, Ding YJ et al (2017) Experimental study on HSS model parameters of soft soil in Shanghai area. Chin J Geotechn Eng 39(02):269–278

Liu C (2008) Study on deformation and internal forces of foundation pit support considering different soil strength/deformation parameters and spatial effects. PhD dissertation, Tianjin University, Tianjin

Liu GB, Wang WD (2009) Foundation pit engineering manual, 2nd edn. China Architecture & Building Press, Beijing

Roscoe KH, Burland JB (1968) On the generalized stress-stain behavior of "wet" clay. In: Engineering plasticity. Cambridge University Press, London, pp 535–609

Roscoe KH, Schofield AN, Thurairajah A (1963) Yielding of clay in states wetter than critical. Geotechnique 13(03):211–240

Schanz T, Vermeer PA, Bonnier PG (1999) The hardening soil model: formulation and verification. In: Beyond 2000 in computational geotechnics, 10 years of PLAXIS international. Proceedings of the international symposium, Amsterdam, pp 281–296

Shanghai Housing and Urban-Rural Construction Management Committee (2018).

Stokoe KH, Darendeli MB, Gilbert RB et al (2004) Development of a new family of normalized modulus reduction and material damping curves. In: Proceedings of NSF/PEER international workshop on uncertainties in nonlinear soil properties and their impact on modeling dynamic soil response, University of California at Berkeley

Vucetic M, Dobry R (1991) Effect of soil plasticity on cyclic response. J Geotechn Eng 117(01):89–107

Wang HR (2012) Research on deformation prediction and environmental impact assessment methods for deep excavations in Shanghai soft soil area. PhD dissertation, Tongji University, Shanghai

Wang RL, Zhang DM (2013) Mechanism and control index of transverse deformation of shield tunnels in soft soil under overload. Chin J Geotechn Eng 35(06):1092–1101

Wang WD, Wang HR, Xu ZH (2012) Experimental study on parameters of hardening soil model for numerical analysis of foundation pit excavation. Rock Soil Mech 33(08):2283–2290

Wang WD, Wang HR, Xu ZH (2013) Study on HS-small model parameters for numerical analysis of foundation pit excavation in Shanghai area. Rock Soil Mech 34(06):1766–1774

Zhou EP (2010) Application of hardening soil model considering small strain in foundation pit deformation analysis. Master's thesis, Harbin Institute of Technology, Harbin

Zhou Q, Shen X, Li XM (2020) Study on transverse deformation characteristics of shield tunnels in soft soil areas. J Zhejiang Univ Water Resour Electr Power 32(01):47–51

Chapter 5
Case Studies and Control Technologies for Loading/Unloading Behavior Over Tunnels

5.1 Introduction

In recent years, with the rapid urban development, numerous engineering activities have been carried out along urban rail transit lines, often impacting the adjacent metro tunnel structures. Among these, significant structural issues have arisen due to engineering loads, such as improper dumping of excavated soil, river channel backfilling, artificial landscaping, and building loads. These issues include tunnel settlement, convergence, misalignment, and severe structural damage such as leakage, segment cracking, and bolt fracture. Additionally, cases of tunnel uplift, leakage at ring and longitudinal joints, and track bed cracking have occurred due to unloading from earth excavation during underground space development. These problems pose serious risks to the structural integrity, durability, and operational safety of the tunnels.

Previous chapters have employed analytical methods, model tests, and numerical simulations to investigate the patterns of longitudinal settlement and transverse convergence deformation in metro tunnels subjected to loading and unloading. However, due to the complexity of tunnel structures and soil properties, all these methods inevitably involve certain assumptions or model simplifications. Consequently, the research findings must be validated through comparative analysis with real engineering cases.

This chapter summarizes five engineering cases with relatively comprehensive deformation records, including two cases of loading above tunnels and three cases of unloading due to foundation pit excavation. The deformation and visible damage characteristics of the underlying tunnels following loading/unloading are analyzed, along with the governance measures implemented and their effectiveness. Numerical simulation results are also used to verify the reliability of the analytical methods.

R. Wang et al., *Impact Analysis and Control Strategies of Overlying Loading/Unloading Effects on Soft Soil Shield Tunnels*, Key Technologies for Tunnel Construction under Complex Geological and Environmental Conditions,
https://doi.org/10.1007/978-981-95-1462-5_5

5.2 Loading Cases and Control Measures

5.2.1 Case Study 1: Surcharge Loading Above Metro Tunnel Section A

(1) Project Overview

In May 2010, large-scale surface earth filling was discovered above the up-track shield tunnel in Section A of a metro line. The maximum fill height reached 7 m, causing severe and concentrated water leakage and structural damage within the tunnel. Some lining rings experienced extreme damage, including cracking of the top segments and bolt fractures, posing significant risks to both structural integrity and operational safety. This section is located in a soft soil area, and the tunnel structure was completed in October 2008, with trial operations commencing in June 2010.

According to the geological data of the shield tunnel section, Layer ⑤ has a relatively large thickness, and the top of Layer ⑦2 is buried deeply. The strata corresponding to the tunnel mainly include Layer ③, Layer ④, and Layer ⑤$_1$ (with partial involvement of the interlayer within Layer ③). In the overloaded area, the shield tunnel lies primarily within Layer ⑤$_1$, overlain by Layer ④ soil, as shown in Fig. 5.1. The basic physical and mechanical parameters of the main soil layers within this section are provided in Table 5.1. The tunnel depth in this area mostly corresponds to mucky clay in Layer ④, which is characterized by high water content and low strength.

(2) Overview of Overloading

According to relevant information, in December 2008, after the completion of the up-track tunnel, an average fill height of approximately 4 m was observed directly above the tunnel section between segment rings 280 and 620. The maximum fill height above ring 570 reached 7 m, with a ratio of surcharge height to overlying soil thickness up to 0.43, indicating a significant overloading magnitude. The surface surcharge conditions are shown in Table 5.2. The surcharge mainly affected the up-track tunnel, with the toe of the soil pile located at the projection edge of the down-track tunnel—i.e., the surcharge did not directly act above the down-track tunnel.

(2) Analysis of the Impact of Surcharge on Tunnel Structure

After the surcharge above the tunnel was identified, the metro maintenance department immediately implemented on-site monitoring programs to observe its impact on the tunnel structure. Based on monitoring data and the observed structural damage and other defects, an analysis of the surcharge's impact on the tunnel was carried out.

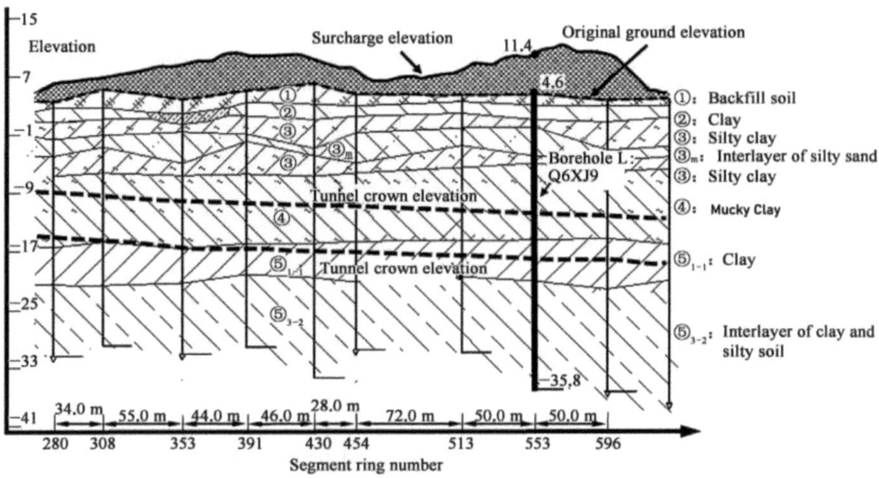

(a) Longitudinal section diagram

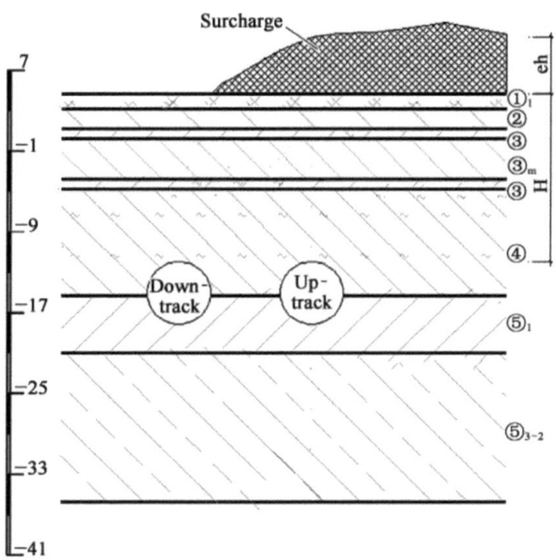

(b) Cross-Section Diagram

Fig. 5.1 Schematic longitudinal and cross-sectional profiles of tunnel geology and surcharge distribution

(1) Monitoring Scheme

Considering the longitudinal influence range of the surcharge on the shield tunnel (Fig. 5.1), monitoring was conducted for tunnel bed settlement, horizontal diameter variations, and opening width of the longitudinal joints from ring 270 to ring 630. The layout of the monitoring points along the cross-section is shown in Fig. 5.2. The joint

Table 5.1 Soil properties

Layer number	Layer name	Unit weight/ (kN·m^{-3})	Water content/ %	Consolidated quick shear		Cone resistance/ MPa	Standard penetration test (SPT) N-value	Sensitivity
				c/kPa	ϕ/(°)			
③	Silty clay	17.5	40.4	12	20	0.55	1.5	4.6
④	Mucky clay	16.6	50.1	14	12.5	0.68	1.6	5.7
⑤$_1$	Clay	17.3	42.3	16	14	0.91	2.7	6.1
⑤$_{3-1}$	Silty clay	18	34.2	17	20	1.74	5.9	

Table 5.2 Correspondence between on-site pile height and tunnel ring number (unit: m)

Ring number	Overlying soil	Surveyed ground elevation	Pile top elevation	Loading height	Elevation after unloading	Pile height/ Original cover depth ratio
280	12.9	3.7	5.5	1.8	4.8	0.14
320	14.8	5.0	7.4	2.4	5.0	0.16
360	14.3	4.3	8.1	3.8	4.5	0.26
400	16.1	3.9	8.7	4.8	3.9	0.30
440	16.5	3.5	7.4	3.9	3.5	0.24
480	15.9	3.4	5.8	2.4	3.4	0.15
520	16.3	3.6	7.5	3.9	3.6	0.24
560	16.5	3.6	9.7	6.1	3.9	0.37
570	16.4	4.4	11.4	7.0	4.4	0.43
600	16.2	4.5	10.2	5.7	4.5	0.35

opening at the tunnel crown was only used to observe the magnitude before unloading. During the subsequent unloading and reinforcement phases, tunnel deformation was mainly analyzed based on tunnel bed settlement and horizontal diameter variations to evaluate the effectiveness of unloading and remediation measures.

(2) Analysis of Tunnel Deformation Monitoring Data

According to the horizontal diameter monitoring data, the deformation pattern of the tunnel structure in the surcharge area was observed and compared with the surcharge height-to-overburden ratio.

Under the influence of the surcharge, the horizontal diameter deviation ΔD of the up-track tunnel [the right tunnel in Fig. 5.1b] significantly increased, showing a typical "horizontal egg-shaped" deformation. Two major convergence zones centered around ring 407 and ring 576 were formed, as shown in Fig. 5.3. In areas with significant overloading, the horizontal diameter variation ΔD was also considerable—at

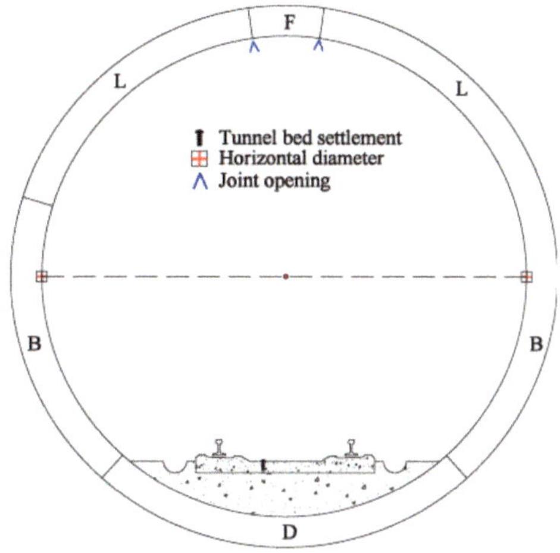

Fig. 5.2 Schematic diagram of monitoring point layout

ring 576, the horizontal diameter increased by 21.4 cm compared to the design value, with a relative ratio of 34.6‰. When comparing the convergence distribution with the settlement distribution, ring 407 was located in a convergence deformation zone with obvious settlement, synchronized with the degree of overloading.

Along the entire line, outside of the surcharge area, the shield tunnel with single-circle through-wall joints had a measured average horizontal diameter deviation of only 3.2 cm in early 2010. However, in the surcharge area, the up-track tunnel between rings 280 and 620 showed an average variation of 14.5 cm, significantly higher than the line average. It is worth noting that the average horizontal diameter deviation of the down-track tunnel in the surcharge zone was only 3.5 cm, falling within the range of normal tunnel deformation. Therefore, it was not included in the subsequent monitoring and remediation efforts.

(3) Tunnel Structural Damage

Under sudden surcharge loading, the tunnel exhibited obvious horizontal egg-shaped deformation. The inner sides of longitudinal joints B–D and F–L opened, while the outer sides of joints B–L opened. The circumferential bolts connecting rings F and L were under high stress, and stress concentration was observed on the inner concrete side of joints B and L. Relatively speaking, the outer waterproof strips were more prone to failure, leading to leakage problems. Based on the on-site structural inspection, severe convergence deformation in the surcharge zone significantly compromised the structural integrity and serviceability of the shield tunnel.

① Segment Structural Damage

Figure 5.4 shows the schematic diagram of the tunnel crown longitudinal joint opening measurement. A vernier caliper was used on-site for measurement. According to the measurement data of the tunnel crown joint opening, a significant joint opening was observed (see Table 5.3). The maximum opening was recorded at segment ring 570, reaching 27.1 mm.

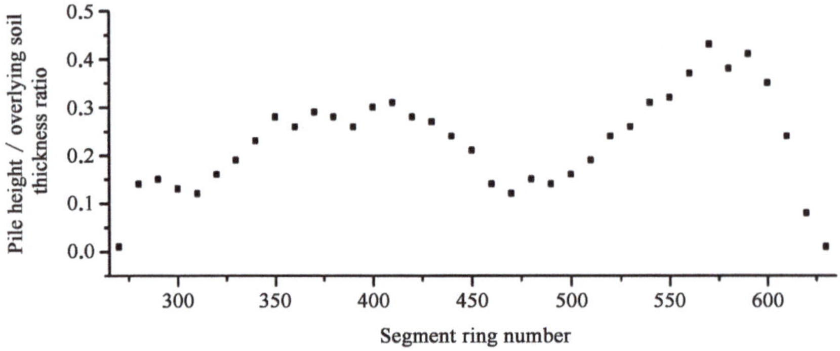

(a) Ratio of Pile Height to Overlying Soil Thickness

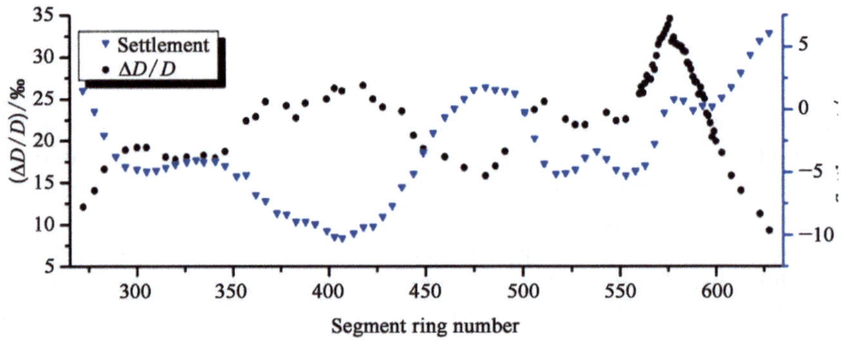

(b) Distribution of Settlement and Horizontal Diameter Variation

Fig. 5.3 Comparison of pile height, settlement, and horizontal diameter variation

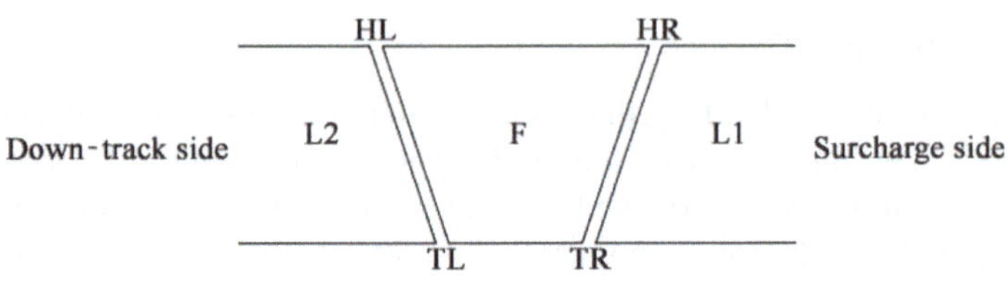

Fig. 5.4 Schematic diagram of tunnel crown longitudinal joint opening measurement

As shown in Fig. 5.5, the crown longitudinal joints in the surcharge zone exhibit noticeable opening, with significant tension observed in the circumferential bolts, and even cases of bolt fracture. Figure 5.6 shows the damaged bolt condition observed at segment ring 581 during on-site inspection. Under the internal force caused by bolt tension and transverse deformation of the segment cross-section, cracks are prone to occur at the segment corners. As shown in Fig. 5.7, after a large portion of concrete cracked and detached at ring 571, it posed a severe threat to structural safety.

Table 5.3 Measured opening widths of tunnel crown longitudinal joints

Ring number	Measured value/mm			
	HL	TL	HR	TR
320	12.9	11.8	15.8	14.0
400	15.9	21.4	11.0	20.1
480	14.5	11.0	16	11.0
570	17.7	27.1	17.3	25.5
620	14.3	13.3	13.8	12.9

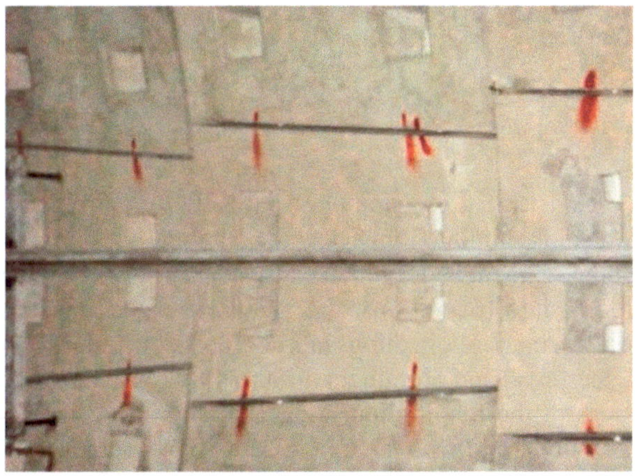

Fig. 5.5 Schematic diagram of crown longitudinal joint opening

Fig. 5.6 Schematic diagram of bolt fracture

Fig. 5.7 Schematic diagram of concrete spalling

Under the horizontal egg-shaped deformation, the concrete on both sides of the waist longitudinal joints acts as rotational supports, resulting in stress concentration at the segment ends, which causes the surface concrete to crush and spall. According to on-site observations, the analysis results are consistent with the actual conditions: concrete crushing and compression failure around the waist joints are both widespread and prominent. Figure 5.8 shows the observed concrete crushing on both sides of the longitudinal joint at ring 570, occurring within approximately 10 cm from the joint, with a damage depth of 1–2 cm.

② Joint Leakage

In shield tunnels undergoing horizontal egg-shaped deformation, the inner sides of B and L longitudinal joints at the waist act as supports, while the outer sides open, which makes the elastic gaskets prone to loosening and failure. On-site observations revealed widespread leakage in the surcharge area, particularly severe at the waist

Fig. 5.8 Schematic diagram of concrete spalling

(a) Waist (b) Crown

Fig. 5.9 Illustration of water leakage at the tunnel waist and crown

longitudinal joints. In some segments, mud and sand seepage was also observed, as shown in Fig. 5.9.

At the tunnel crown and invert, the outer sides of the longitudinal joints act as supports, and the inner sides open, making waterproofing components relatively less susceptible to failure. However, due to compressive stress acting on the outer crown end, localized concrete crushing can occur. If crushing cracks appear on the outer side, resulting in water leakage, it could pose a very serious problem.

Based on the preliminary analysis of tunnel structural damage induced by the surcharge, the longitudinal joints are the most severely affected and densely damaged zones. The main defect is leakage, concentrated at longitudinal joints. There are clear signs of crown joint opening and concrete crushing on both sides of the waist joints. In some rings, the crown concrete is visibly cracked or spalled. No significant opening or misalignment was observed at the circumferential joints, indicating that the longitudinal bolts at ring joints are functioning properly.

(3) Tunnel Remediation Procedure

Based on the previously analyzed structural damage, the water leakage and structural deterioration resulting from transverse convergence deformation in the surcharge zone of the shield tunnel pose a serious threat to the structural integrity and operational safety of the tunnel. In contrast, the inter-ring misalignment or joint opening caused by longitudinal settlement of the shield tunnel has not led to significant serviceability issues and does not compromise structural safety. Therefore, the remediation efforts primarily focus on addressing cross-sectional convergence deformation. The detailed remediation process is illustrated in Fig. 5.10.

Since the tunnel defects were caused by surcharge loading from above, the first measure taken was to organize immediate unloading of the overlying soil to minimize the duration of overloading acting on the tunnel structure. Subsequently, polyurethane

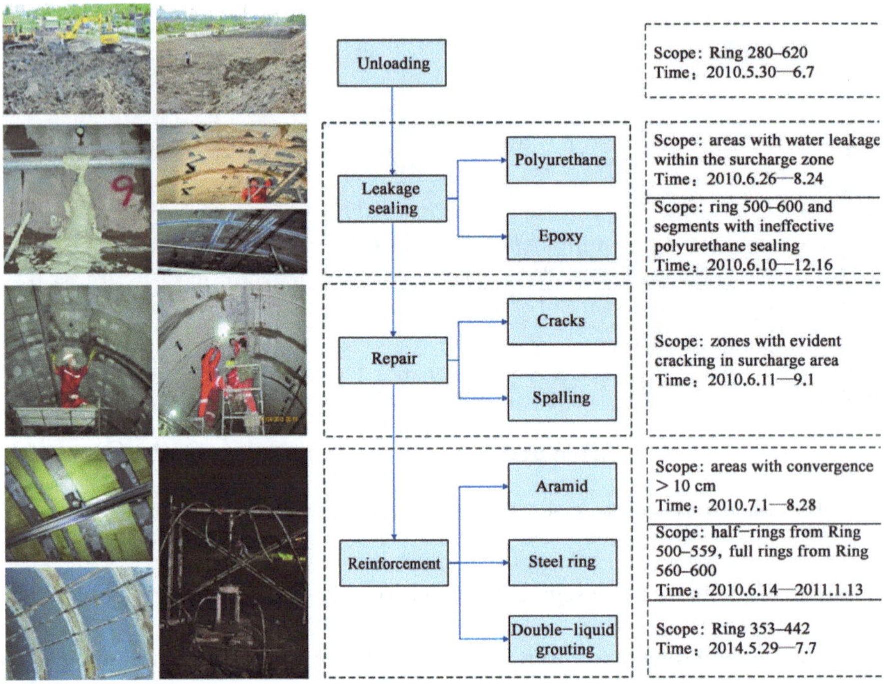

Fig. 5.10 Schematic diagram of tunnel remediation process

and epoxy were applied promptly to seal the water leakage in the tunnel, in order to prevent further deterioration of the structure, avoid aggravated lateral pressure loss, and suppress the vicious cycle of transverse deformation.

Prior to applying aramid fiber sheets, damaged concrete was removed, cracks were repaired, and surface treatment of the segment structure was carried out to maintain structural integrity. Afterwards, structural reinforcement was implemented to enhance the load-bearing capacity of the tunnel structure. Aramid fiber sheets were first applied for temporary strengthening. For the severely deformed transverse convergence areas between rings 500–559 and rings 560–600 of the up-track tunnel, steel ring reinforcement was applied in half-ring and full-ring configurations, respectively.

(1) Leakage Sealing

① Polyurethane Sealing

The water leakage in the tunnel was primarily observed at the segment joints. A behind-the-lining injection method using water-soluble polyurethane was adopted on-site for sealing. The water-soluble polyurethane expands and foams upon contact with water, forming a waterproof membrane on the outer surface of the segments. This blocks water sources from entering the tunnel through joints, or significantly increases the seepage path, thereby alleviating the leakage-related defects.

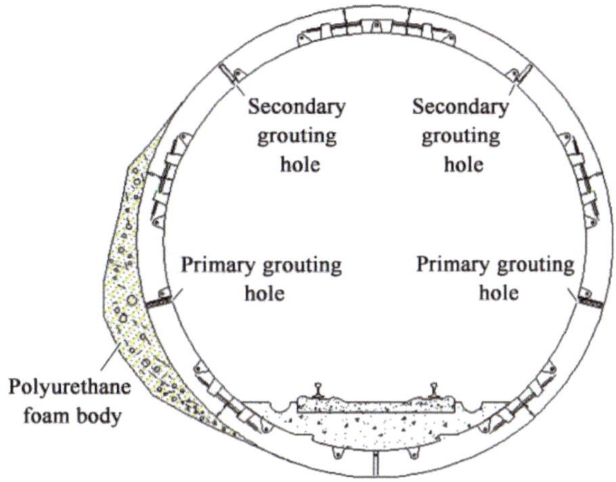

Fig. 5.11 Schematic diagram of polyurethane grouting effect

As shown in Fig. 5.11, polyurethane grouting was mainly carried out through the standard B segment grouting holes on both sides. In a few areas with large and severe leakage, grouting holes in adjacent L segments were also used. The polyurethane grout was injected symmetrically from both sides under pressure.

② Epoxy Sealing

In the steel ring reinforcement areas, to prevent recurring backside leakage and subsequent corrosion of steel plates after the ring assembly was completed, epoxy resin was used to seal the segment joints. In areas where leakage was severe and polyurethane sealing was ineffective, epoxy grouting was also employed.

Figure 5.12 illustrates the epoxy grouting scheme, in which dual-ring or multi-ring joints were selected as sealing units. Taking the dual-ring configuration as an example, below the crown block, elastic epoxy was injected from bottom to top to fill and seal the joint zone enclosed by the grouting boundary. Above the crown block, due to structural integrity requirements, rigid epoxy was injected from bottom to top to seal the enclosed joint area and form a watertight curtain along the joints.

(2) Reinforcement

① Aramid Fiber Sheet Application

Figure 5.13 shows the on-site application of aramid fiber sheets for reinforcing the tunnel crown structure. The specific reinforcement area covers approximately 58° at the crown, with a circumferential arc length of about 2.8 meters. The aramid fiber sheets were applied using rigid epoxy resin with high bonding strength. For each ring, the aramid sheets were divided into three sections: the middle section corresponds to the area between the hand holes, and the other two strips were applied near the circumferential joints on both sides of the segments.

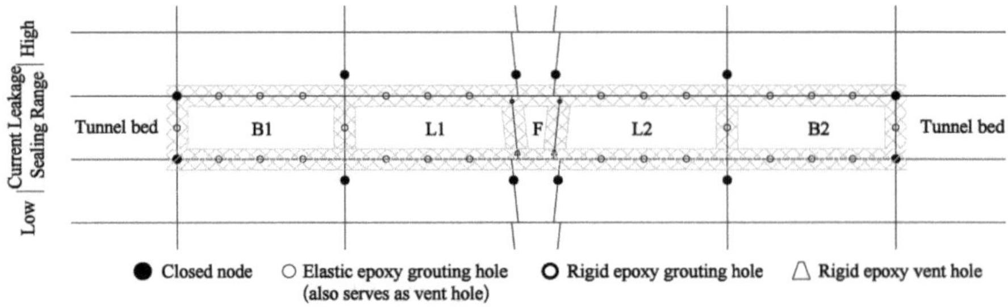

Fig. 5.12 Schematic diagram of epoxy grouting hole arrangement

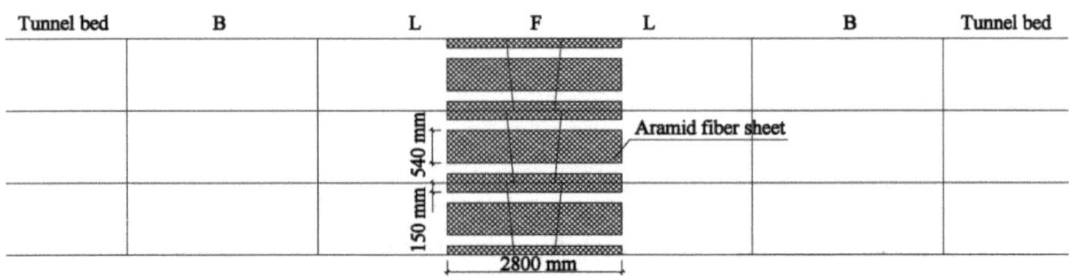

Fig. 5.13 Schematic diagram of aramid fiber sheet application

② Steel Ring Reinforcement

Figure 5.14 shows the schematic diagram of on-site steel ring reinforcement. The full-ring reinforcement consists of three steel segment plates, two steel segment plates with corbels, and one steel tension plate at the tunnel bed, forming a closed load-bearing ring system to enhance the deformed shield tunnel's capacity to resist eccentric loading. The procedure involves first installing the corbels, then assembling the three upper steel segments, and finally connecting both corbels using the steel tension plate.

The half-ring reinforcement consists of two steel segment plates with corbels and one steel tension plate at the tunnel bed.

(3) Double-Liquid Grouting

This method uses a mixture of cement grout and sodium silicate, which, once injected into the soil, forms a reinforcement body. The sodium silicate rapidly solidifies the cement slurry, enabling the grouting body to gain strength in a short period. This not only minimizes disturbance to surrounding soil structures but also satisfies the short-term structural requirements of the tunnel. In summary, double-liquid grouting achieves both ground reinforcement and disturbance control.

The cement grout has a water–cement ratio of 0.6–0.7. The sodium silicate has a modulus of 2.85 and Baumé degree of 35, with a mixing ratio of 3:1 (cement grout to sodium silicate).

The grouting schematic is shown in Fig. 5.15. The reinforcement depth is within 5 m above the tunnel invert. One row of grouting holes was arranged between the

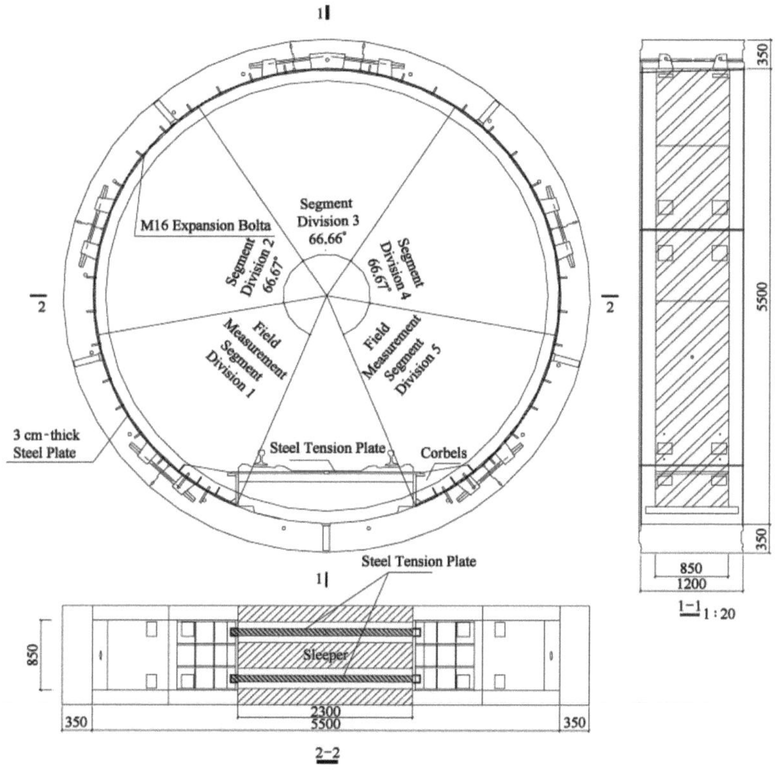

Fig. 5.14 Schematic diagram of steel ring reinforcement

up-track and down-track tunnels, and three additional rows were placed on the outer side of the up-track tunnel. The longitudinal spacing between grouting holes was 1.2 m, and the transverse spacing was 0.6 m.

(4) Remediation Effectiveness

During the remediation process, the time history of tunnel horizontal convergence and settlement was observed to assess the effectiveness of each remediation stage. Since different types of remediation measures overlapped in time, the entire remediation period from May 2010 to July 2014, when double-liquid grouting was completed, was divided into three main stages. Table 5.4 summarizes the time-dependent settlement and convergence variations at ring 407 during each stage.

Key observations during unloading and grouting include: During the unloading phase, the maximum uplift at ring 407 of the up-track tunnel reached 14.1 mm, resulting in a cumulative heave state. However, the convergence recovery was only 6 mm. Compared to longitudinal settlement, unloading was not particularly effective in restoring convergence deformation, indicating that radial deformation has relatively poor recoverability. This suggests that overloading has a more severe and

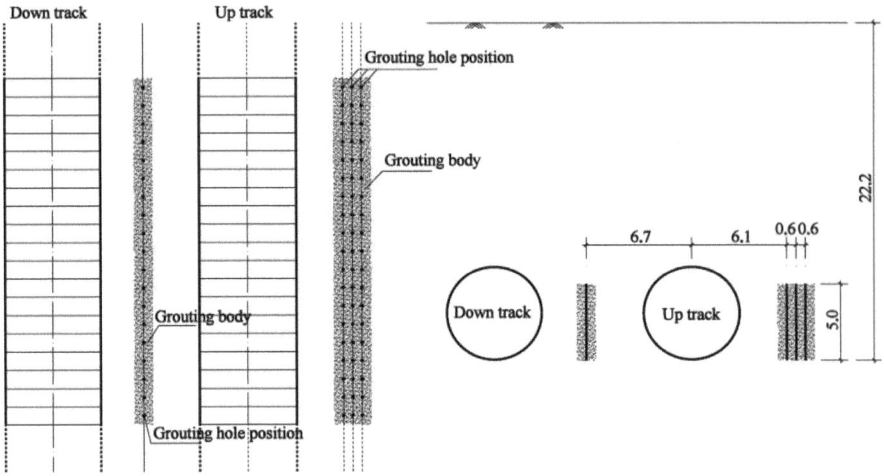

Fig. 5.15 Schematic diagram of double-liquid grouting

Table 5.4 Remediation effect statistics

Remediation stage		Time span	Settlement change/mm		Convergence change/cm	
Stage 1	Unloading	7 days	14.1	14.1	−0.6	−0.6
Stage 2	Leak sealing	8 months	8.6	13.6	−0.2	−0.5
	Repairing		10.9		−0.3	
	Aramid		2.6		−0.2	
	Steel ring		10.2		−0.5	
Stage 3	Grouting	1 month 8 days	7.6	7.6	−3.9	−3.9

Note Steel ring reinforcement was not applied to Ring 407, but according to monitoring data, the steel ring reinforcement construction itself has a minimal impact on the tunnel structure at the corresponding location

persistent impact on tunnel cross-sectional deformation, which is more challenging to remediate.

In the second stage, leakage sealing, crack repair, and reinforcement measures (such as aramid and steel rings) were implemented over a period of more than eight months. During this time, the tunnel experienced a further settlement reduction of 13.6 mm and a convergence recovery of 5 mm. These measures did not directly improve longitudinal settlement or radial convergence; rather, they continued the unloading effects that had already taken place. Historical settlement data indicates that most recovery occurred within the first month of the stage.

In the third stage, sidewall grouting reinforcement of the tunnel began in May 2014. The grouting significantly improved the tunnel's convergence condition: horizontal diameter at ring 407 shrank by nearly 4 cm during the grouting period.

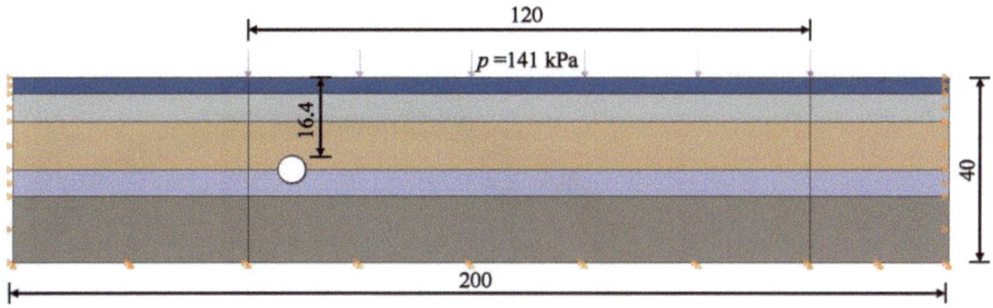

Fig. 5.16 Finite element model diagram (unit: m)

However, within two months after grouting, the diameter increased by 7 mm again. During the grouting process, the grouting pressure and volume caused a short-term increase in pore water pressure around the tunnel sides, which compressed the tunnel's transverse deformation. Once the grouting ended, pore water pressure gradually dissipated, and part of the initial compression effect was lost.

(4) Numerical Simulation Analysis

Using the finite element analysis method for shield tunnel transverse deformation under top loading described in Chap. 3, an ABAQUS finite element numerical model was developed for this case study. Figure 5.16 shows the finite element model. The simulation focused on ring 570, where deformation was most severe. The tunnel depth and ground conditions were modeled based on the actual conditions of this ring. The surface loading width was set at 120 m, and the applied load was 141 kPa, corresponding to a pile height-to-cover thickness ratio of 0.43.

Figure 5.17 illustrates the transverse deformation of the tunnel before and after loading (with deformation magnified by a factor of 5 for clarity). Under the applied loading, the tunnel ring exhibited horizontal displacement to the left, along with significant convergence deformation. Joint openings were evident in multiple areas, resulting in severe deformation and structural damage to the tunnel.

According to the finite element model calculations, under surface loading, the horizontal displacement of the left springline of the tunnel was 180 mm, and the horizontal displacement of the right springline was 29 mm, resulting in a horizontal convergence deformation of 151 mm (corresponding to 24.4‰D). These results align closely with measured data.

(5) Summary and Recommendations

Based on the analysis of the effects of large-area overloading on tunnel deformation and the applied remediation methods, the following conclusions and recommendations can be made:

(1) The impact of large-area overloading above the tunnel is significant. Under local overburden heights approaching 7 m, the tunnel exhibited pronounced convergence deformation, with the maximum horizontal diameter change

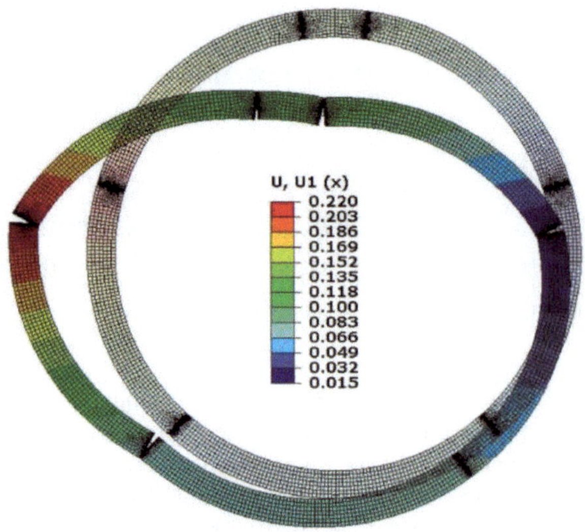

Fig. 5.17 Illustration of tunnel transverse deformation before and after loading

approaching 21 cm, leading to various types of defects associated with convergence deformation. These defects severely affected tunnel operations and structural safety. However, longitudinal differential settlement induced by overloading, as inferred from measured data, did not result in serviceability or structural safety issues.

(2) Unloading was effective in reducing the longitudinal differential settlement trend. After unloading, the rebound reached 14.1 mm, while convergence recovery was minimal at only 0.6 cm (less than 1‰D). Subsequent leakage sealing, crack repair, and aramid-steel ring reinforcement methods did not significantly lift the tunnel or alleviate convergence deformation. The upward trend in the tunnel's alignment was attributed to the lingering effects of unloading, but convergence deformation showed no substantial improvement. However, double-liquid side grouting reinforcement caused significant convergence recovery, with ring 407 experiencing a cumulative convergence shrinkage of approximately 3.9 cm after grouting was completed.

(3) Recommendations. After unloading, leakage remediation should be prioritized to prevent ongoing sidewall leakage, which can lead to further transverse convergence deformation. Subsequent efforts should focus on double-liquid side grouting reinforcement to reduce convergence deformation, increase lateral constraint, and improve the tunnel's structural stress conditions. Once the reinforcement has stabilized, steel ring reinforcement should be implemented.

5.2.2 *Case 2: A Road Loading Case Above Metro Tunnel Section B*

(1) Project Overview and Overloading Condition

(1) Project Overview

On August 22, 2020, analysis of monitoring data from the tunnel revealed abnormal convergence deformation in the under-construction metro line's Section B tunnel. The convergence deformation exceeded 6 cm compared to the theoretical design value, with a maximum measured deformation of 15.9 cm. On-site investigations found significant water leakage inside the tunnel, as well as structural damage such as fractured segment joints, posing a serious threat to the tunnel's structural integrity and the critical deadline for opening the line.

Geological data for the shield tunnel (Fig. 5.18) indicated that the metro tunnel passed through Layer ③ silty clay and Layer ④ mucky clay, underlain by Layer ⑥ silty clay (locally underlain by Layer $⑤_{11}$ clay). The tunnel crown was buried at a depth of 7.5–10 m. The shield tunnel was assembled from six segments with an external diameter of $\phi 6.2$ m and an internal diameter of $\phi 5.5$ m. The basic physical and mechanical parameters of the main soil layers in this section are shown in Fig. 5.19.

(2) Overloading Overview

After abnormal convergence deformation was detected in the tunnel section, the subway maintenance department immediately conducted a site survey along with the relevant units. Upon inspection, it was found that the tunnel damage section corresponds to an intersection location of a municipal engineering project above the

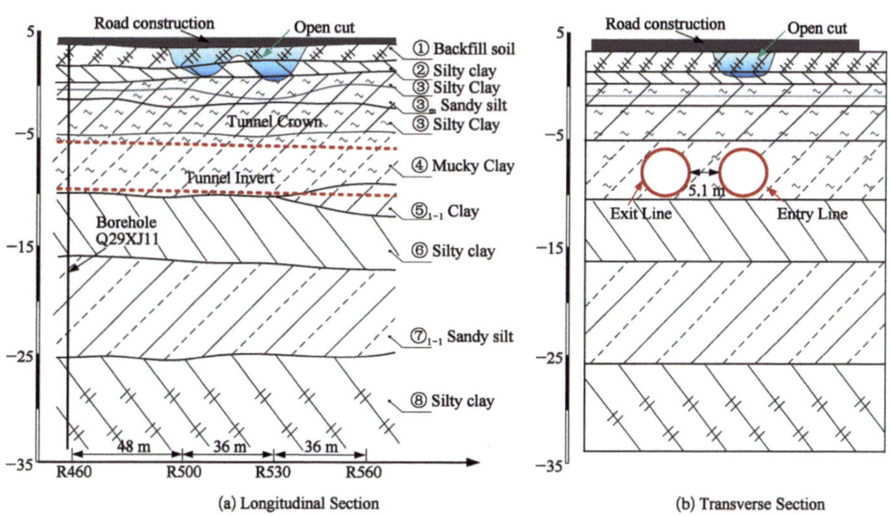

Fig. 5.18 Schematic of tunnel geology and load profile (longitudinal and transverse sections)

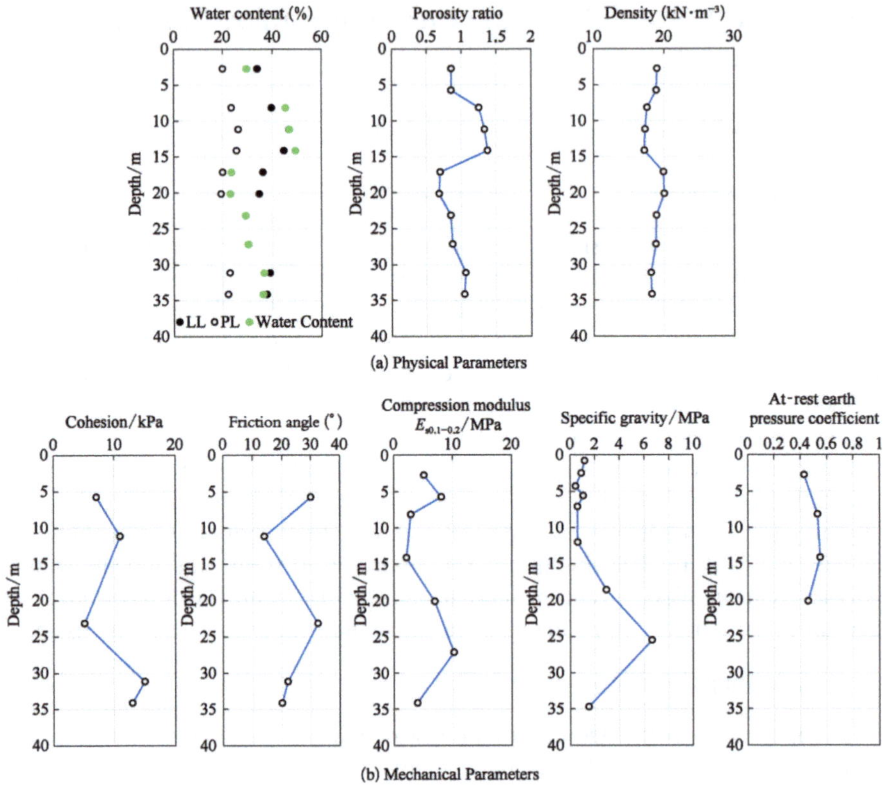

Fig. 5.19 Physical and mechanical parameters of main soil layers in tunnel boring machine (TBM) section

ground. In the past two years, works such as open cut backfilling, new roadbed foundation raising, and pavement laying had been carried out. As shown in Fig. 5.20, the newly constructed road is directly above the subway tunnel, with the road surface design elevation ranging from 4.63 to 6.75 m, and the fill height varying from 0.1 to 2.43 m. Additionally, two open cuts are present in this section, with widths nearly identical to the subway tunnel structure. The length of these cuts is distributed between R460 (K1 + 080) and R560 (K1 + 180), with the direction almost coinciding with the subway tunnel. The water depth of open cut #2 is approximately 0.3 m, with a silt depth of about 1.5 m and a silt bottom elevation of around 1.5 m. Open cut #3 has a water depth of about 1.7 m, a silt depth of approximately 1 m, and a silt bottom elevation of about 0.5 m. The large-scale, substantial, and continuous loading construction, along with the open cut backfilling, is the direct cause of the significant deformation and damage to the subway tunnel beneath.

(2) Analysis of the Impact of Ground Overload on Tunnel Structure

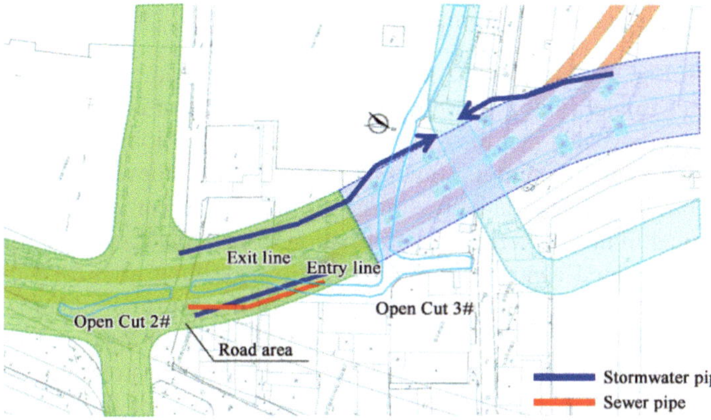

Fig. 5.20 Distribution of roads above the subway tunnel

After the discovery of significant convergence deformation in the tunnel, the monitoring and inspection of the tunnel were intensified. The monitoring points for tunnel convergence deformation are shown in Fig. 5.21. To better capture the structural deformation characteristics during the subsequent tunnel deformation remediation process, a combination of automated monitoring at key sections and manual measurements was used to monitor the tunnel convergence deformation.

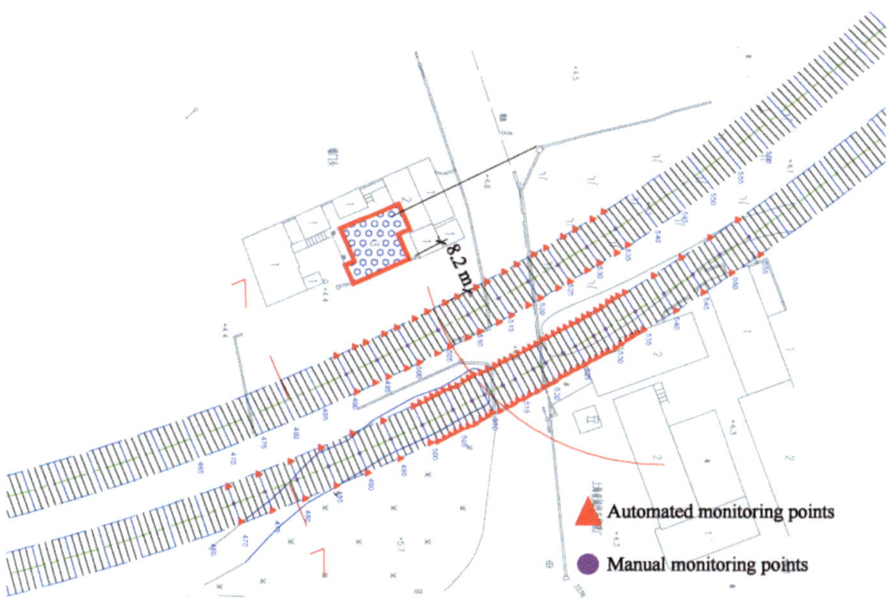

Fig. 5.21 Layout plan of tunnel convergence deformation monitoring points

(1) Tunnel Convergence Deformation and Joint Opening

As shown in Fig. 5.22, under the influence of surface road construction, both the exit and entry tunnel lines experienced varying degrees of convergence deformation. Inspection revealed that the top segment longitudinal joints were significantly opened, and the joint opening is shown in Fig. 5.23. The convergence deformation in the exit line tunnel is relatively similar to that in the entry line tunnel, and is nearly identical to the convergence deformation in the tunnel structure of the entry line, which is located away from the open cut backfilling area (6–8 cm). It can be concluded that the convergence deformation of the exit line tunnel is primarily caused by the road construction directly above. Early monitoring data indicated that the convergence deformation of the tunnel before road construction was approximately 3 cm. Compared to the convergence deformation of the exit line tunnel, the increment in tunnel convergence deformation caused by road construction was approximately 3–5 cm.

After the surcharge load was applied, to further understand the deformation of the tunnel structure, a caliper was used on-site to measure the joint opening between the top slab (F) and two connecting blocks (L1 and L2), as shown in Fig. 5.23. For the entry line tunnel, within the ring sections R500-R530, the joint opening between F-L2 was significantly larger than the joint opening between F-L1. In the ring sections R490-R500 and R530-R540, the joint openings of both connections were nearly the same. The main reason for this is that there is a certain deviation between the two open cuts above the entry line tunnel and the tunnel, which leads to an uneven load distribution in the entry line tunnel, causing asymmetric deformation in the tunnel cross-section. Compared to the entry line tunnel, the deformation of the exit line tunnel is primarily caused by road construction. The additional load acting on the tunnel is symmetrically distributed, leading to a symmetric deformation pattern in the tunnel cross-section, and thus, the joint openings of both connections are nearly the same, with an average opening of 12 mm.

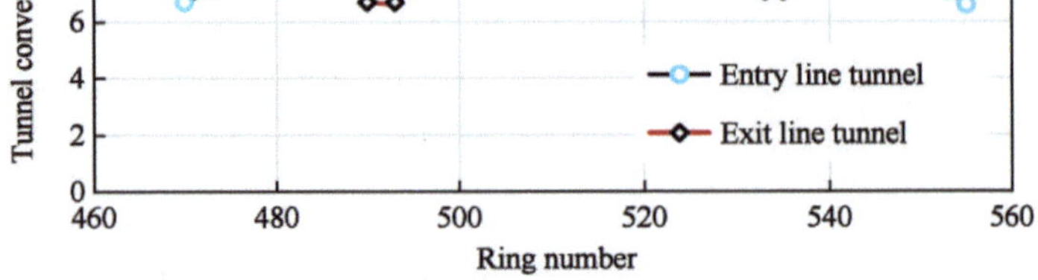

Fig. 5.22 Tunnel convergence deformation

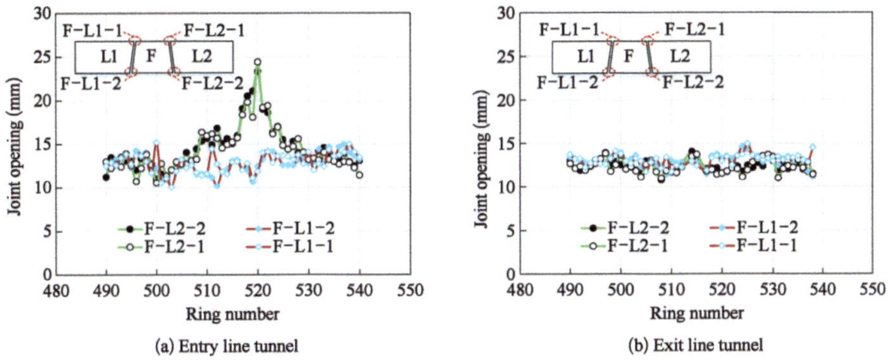

Fig. 5.23 Joint opening measurement

(2) Structural Damage Induced by Road Backfilling

Under the effect of additional loads, the tunnel structure exhibits a distinct "horizontal duck egg" deformation. The inner surface of the top and bottom joints opens, and the outer surface of the shoulder joints opens, resulting in higher stress on the top joint bolts and significant stress concentration on the inner surface of the shoulder joint concrete. This makes the shoulder joint more prone to seal failure, leading to water and slurry leakage.

Water leakage inside the tunnel typically manifests as joint leakage (Fig. 5.24). Table 5.5 shows the field statistics of internal water leakage in the segment. From Table 5.5, it can be observed that surface road construction caused significant water leakage in the entry line tunnel between R475 and R550, while the exit line tunnel between C495 and C500 showed damp spots. The reason for this is that the open cut backfilling caused excessive convergence deformation in the entry line tunnel, leading to a significant opening of the joints. Once the water-barrier strips and seals fail, leakage passages are formed. By combining the tunnel convergence deformation and the leakage patterns, it can be seen that when the tunnel convergence deformation exceeds 7 cm, joint leakage begins to occur.

During the development of the tunnel's transverse deformation, cracks or even spalling of the concrete at the joint corners are likely to occur. In the entry line tunnel, the convergence deformation of ring R520 reached 15.9 cm, and the top joint exhibited a joint opening of up to 25 mm, causing noticeable cracks in the concrete at the joint corner (Fig. 5.24).

(3) Tunnel Large Deformation Remediation Measures

Based on the convergence deformation and damage statistics of the tunnel structure from the previous section, it is evident that the surcharge load has caused a series of structural damages, including concrete crushing and water leakage in the segments, which severely threaten the tunnel structure and operational safety. Therefore, the tunnel structural remediation primarily focuses on the convergence deformation of the cross-section. The specific remediation measures include leakage sealing, lateral

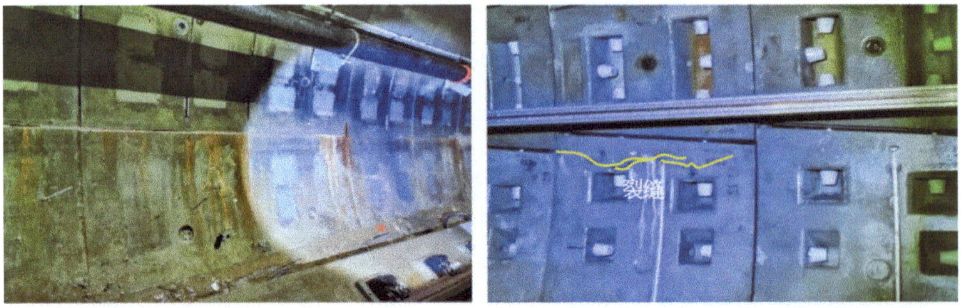

(a) Segment Water Leakage (b)Joint Cracking

Fig. 5.24 Segment damage

Table 5.5 Statistics of water leakage in tunnel sections

Ring number	Convergence value/cm	Water leakage condition	Ring number	Convergence value/cm	Water leakage condition
R475	7.3	Significant water leakage	R535	9.6	Significant water leakage
R480	7.3		R540	9	
R485	8		R545	8.1	
R490	8.9		R550	7.4	
R495	8.9		C495	7.1	Damp spots
R500	9.7		C500	8.9	
R505	11.5		C505	8.1	
R510	13.4		C510	8	
R515	13.3		C515	7.9	
R520	15.9		C520	8	
R525	11.7		C525	8	
R530	10.7		C530	7.6	

micro-disturbance grouting, surface road unloading and backfilling, and steel plate reinforcement. Leakage sealing and steel plate reinforcement are consistent with Case 1, so they will not be elaborated further.

The grouting schematic is shown in Fig. 5.25. Due to the 5.1 m distance between the exit and entry tunnel lines, and the larger convergence deformation size and range in the entry line tunnel compared to the exit line tunnel, grouting was carried out at the following locations: two rows of grouting at the outer side of the C495-C535 ring, 4.2 and 3.6 m positions; four rows of grouting at the outer side of the R455-R560 rings, at 4.2 and 3.6 m positions, and the inner side at 2.4 and 2.7 m positions. A total of 506 grouting holes were planned. The grouting depth was within 5.2 m above the tunnel bottom, with a longitudinal spacing of 1.2 m and lateral spacing of 0.6 m. The remediation project began on August 25, 2020, and grouting started on

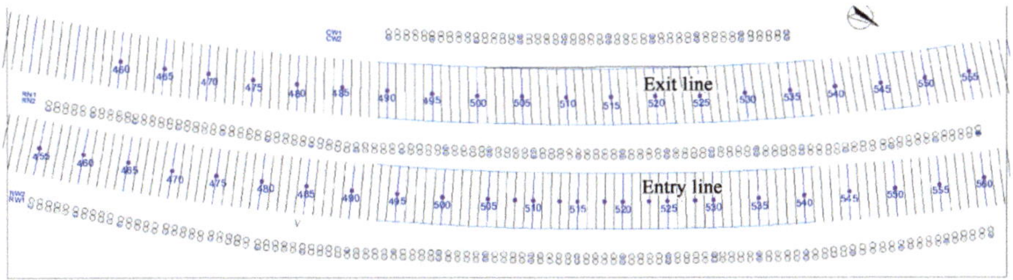

(a) Grouting Hole Location Diagram

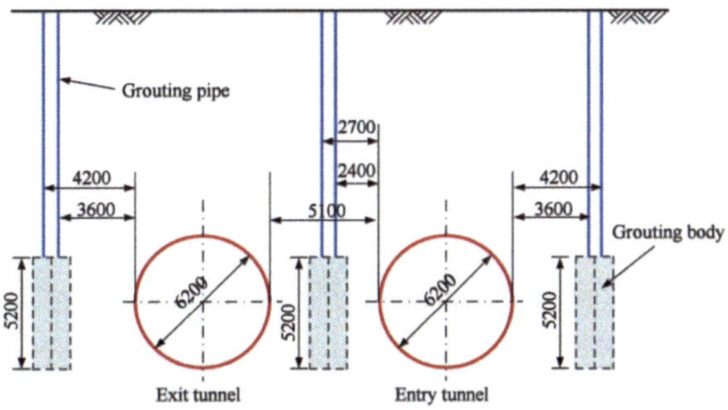

(b) Grouting Cross-Section Diagram (Unit: mm)

Fig. 5.25 Grouting schematic

August 26, 2020, finishing at dawn on December 10, with a total of 353 grouting holes completed.

The main reason for the significant convergence deformation in the tunnel is the presence of two open cuts directly above the entry line tunnel, as well as the road construction and extensive backfilling of the open cuts (Fig. 5.22). The road construction above the tunnel is part of a municipal project, so it is not feasible to directly unload to the initial ground elevation. In response to the actual needs of the project, EPS lightweight material was used to backfill the road construction area directly above the tunnel, thereby reducing the surface load, decreasing the additional load acting on the top of the tunnel, and mitigating the impact of road construction on the tunnel's deformation. The backfilling plan for road unloading and the field construction diagram are shown in Figs. 5.26 and 5.27, respectively. The backfilling process involves over-excavation of the surface, and the main principles for determining the over-excavation depth are: ① Ensuring that the road design height remains unchanged; ② The road backfilling and open cut backfilling will not generate new additional loads.

(4) During the tunnel deformation repair process, vertical displacement, horizontal displacement, convergence deformation, and joint opening measurements for the entry line tunnels between rings 470–580 and 490–535 were taken using

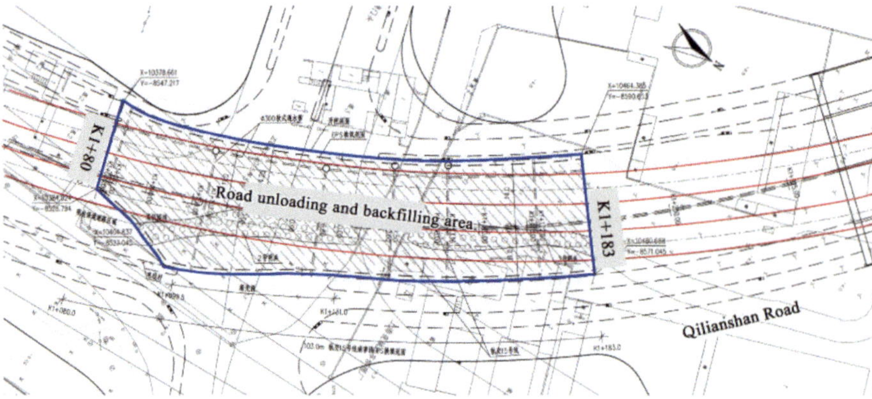

Fig. 5.26 Road unloading and backfilling plan

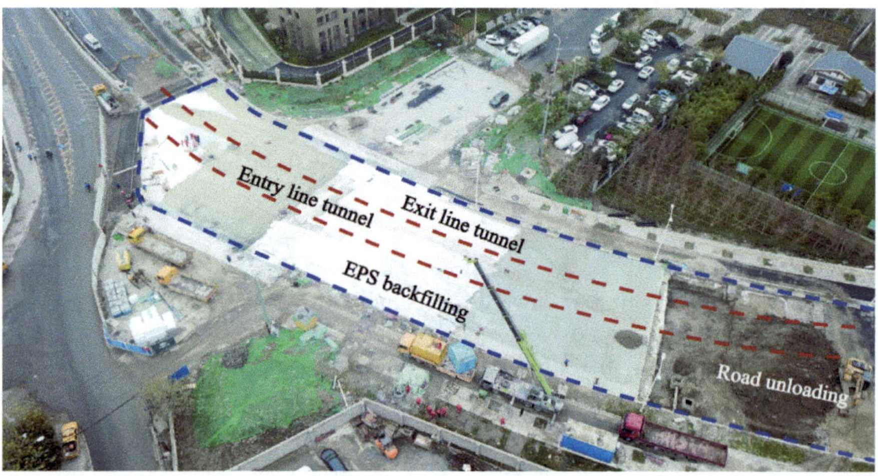

Fig. 5.27 Road unloading and backfilling construction site

both manual and automated monitoring. The monitoring points are shown in Fig. 5.25.

The actual grouting hole locations, grouting times, and corresponding segment ring numbers are shown in Fig. 5.28. Comparing Figs. 5.28 and 5.25, the number of actual grouting holes was smaller than initially planned. During the on-site grouting process, the grouting density for rings 500–530 was significantly higher than for other segment rings, and the grouting started earlier. The grouting sequence was generally RW1/RN1 → CW2 → RW2/RN2.

During the grouting process, certain horizontal displacements in the segment rings were observed. Excessive horizontal displacement could lead to transverse offset deformations in the segment rings, which would trigger joint water leakage.

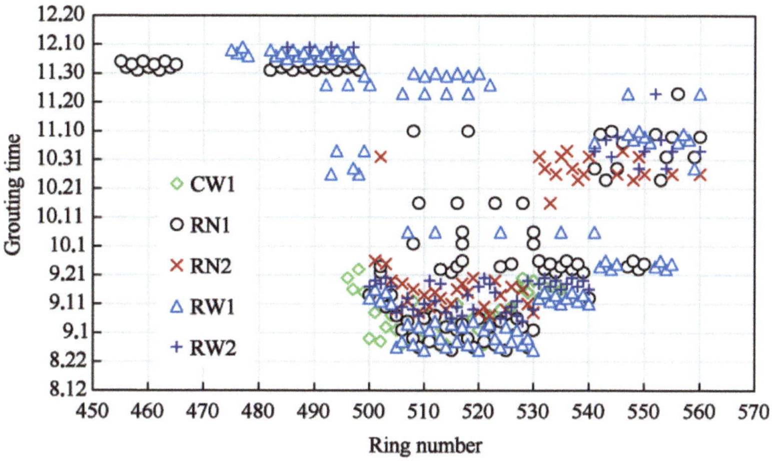

Fig. 5.28 On-site grouting plan

To avoid this issue, vertical grouting was performed using the skip-hole method, and the grouting followed the principles of "uniform, small quantity, multiple points, and repeated" controlled grouting.

The vertical and horizontal displacements of the tunnel during the grouting process are shown in Fig. 5.29. From Fig. 5.29a, it can be seen that the tunnel experienced some degree of heave deformation, with the maximum heave amount being approximately 11 mm, occurring in the entry line tunnel between rings R510 and R520. Since the grouting process was accompanied by road unloading, the tunnel's heave deformation is the result of the combined effects of grouting and unloading. From Fig. 5.29b, it can be observed that the horizontal displacement of the entry line tunnel during grouting was greater than that of the exit line tunnel. However, the average deformation amounts were approximately 2 and 1 mm, respectively, and the maximum lateral offset was less than 2 mm, indicating that the effect on the tunnel's lateral horizontal displacement was minimal. This suggests that the grouting arrangement was reasonable.

The convergence deformation of the tunnel before and after grouting is shown in Fig. 5.30. The improvement in tunnel convergence deformation before and after grouting refers to the difference in convergence deformation before and after grouting. Overall, the maximum improvement in convergence deformation for the entry line tunnel is 7.2 cm, occurring at ring R525. The maximum improvement in convergence deformation for the exit line tunnel is 5.9 cm, occurring at ring C510. Due to the higher grouting hole density in the 500–530 ring section, the convergence deformation improvement in this area is significantly greater than in other areas.

During the grouting process, the change in tunnel convergence deformation over time is shown in Fig. 5.31. In the section between rings R505-R535 (Fig. 5.31a), the convergence deformation initially increased and then decreased. During the periods from August 30 to September 2 and from September 26 to October 1, the grouting

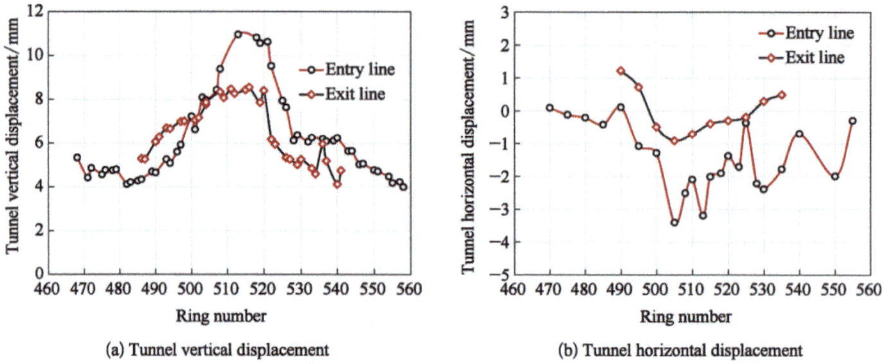

Fig. 5.29 Tunnel displacement during remediation

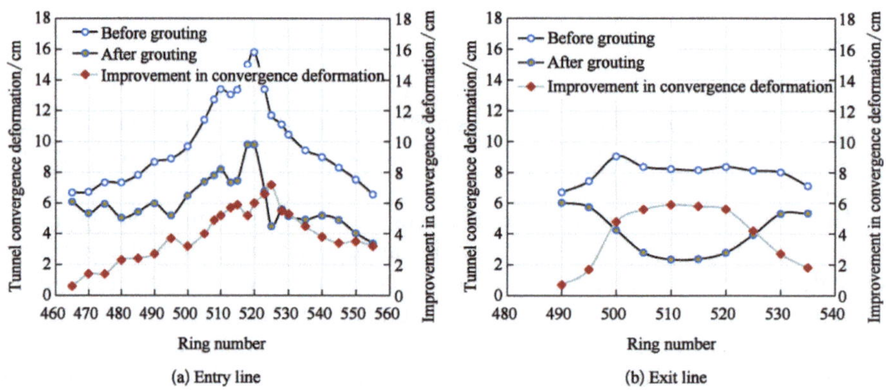

Fig. 5.30 Tunnel convergence deformation before and after grouting

work was temporarily stopped, resulting in a decrease in the change of convergence deformation. After November 29, several previously completed grouting holes were re-grouted, leading to an increase in convergence deformation. In the sections between rings R470-R500 and R540-R555, the grouting points were relatively few, so the change in convergence deformation was significantly smaller compared to the R505-R535 section. Comparing the changes in convergence deformation between these two sections shows that the amount of grouting has a clear impact on convergence deformation. The change in convergence deformation of the exit line tunnel follows a similar trend to the entry line and will not be discussed further.

Figure 5.32 shows the variation of joint opening values at the top slab before and after grouting with respect to the segment ring number. Negative values indicate that the joint is tending towards closure. Grouting results in a reduction in tunnel convergence deformation, and the joint opening of the top slab also decreases. In certain local areas, the change in joint opening for the entry line tunnel is slightly

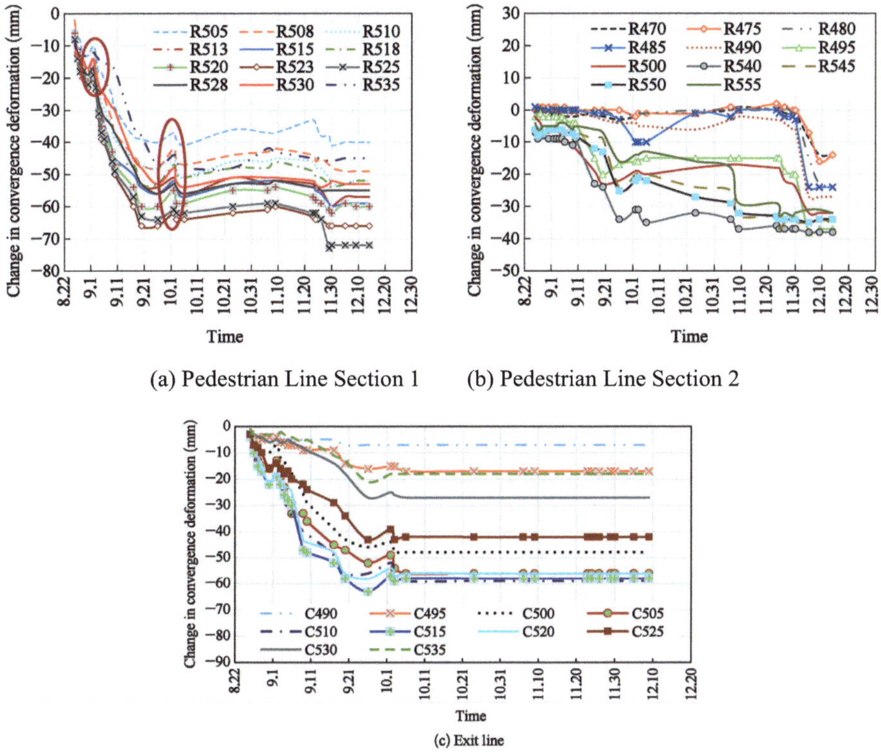

Fig. 5.31 Tunnel convergence deformation change curves during grouting

greater than for the exit line tunnel. Across the entire monitoring area, the change in joint opening values ranges from −1 to −4 mm.

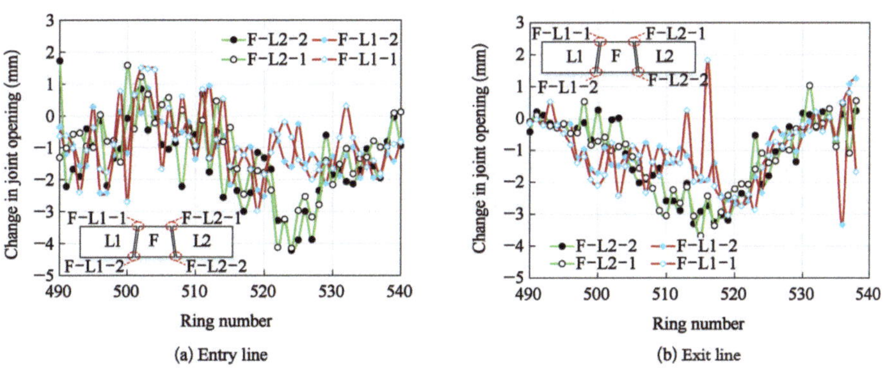

Fig. 5.32 Joint opening during grouting

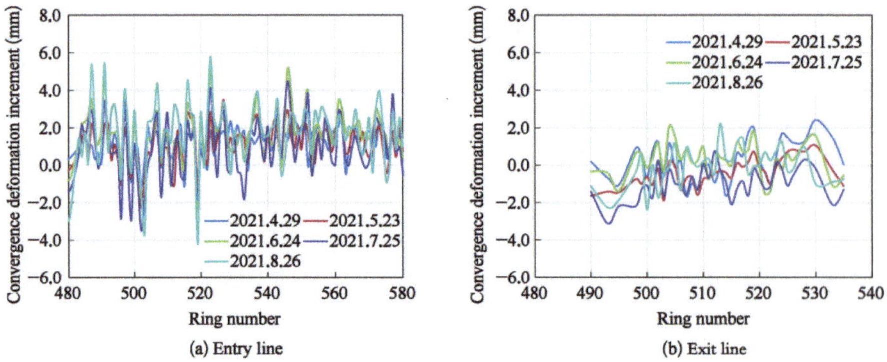

Fig. 5.33 Tunnel convergence deformation after steel plate reinforcement

After the steel plate reinforcement, the tunnel convergence deformation monitoring points were re-established on March 29, 2021, with the convergence deformation on that date serving as the initial value. The deformation increment is shown in Fig. 5.33. In the steel plate reinforcement section, the convergence deformation increments of both the entry line and exit line tunnels did not change significantly, indicating that the tunnel deformation has stabilized after the steel plate reinforcement.

(5) Numerical Simulation Analysis

The numerical analysis method for shield tunnel deformation under surcharge load, introduced in Chap. 3, is used. A numerical model is established using the ABAQUS finite element software, as shown in Fig. 5.34. To simplify the numerical model, the open cut backfilling is directly simulated by applying an equivalent load. The soil mechanical parameters are simulated using the Mohr–Coulomb model, with the relevant parameter values provided in Fig. 5.19. The tunnel segments are modeled using solid elements, and the bolts are modeled using nonlinear spring elements that only experience tension.

Under road construction conditions, the convergence deformation of the entry line and exit line tunnels is shown in Fig. 5.35. The numerical simulation results indicate that the convergence deformation of the entry line tunnel is approximately 12 cm, and the convergence deformation of the exit line tunnel is approximately 7.8 cm. The tunnel convergence deformation is in good agreement with the measured results.

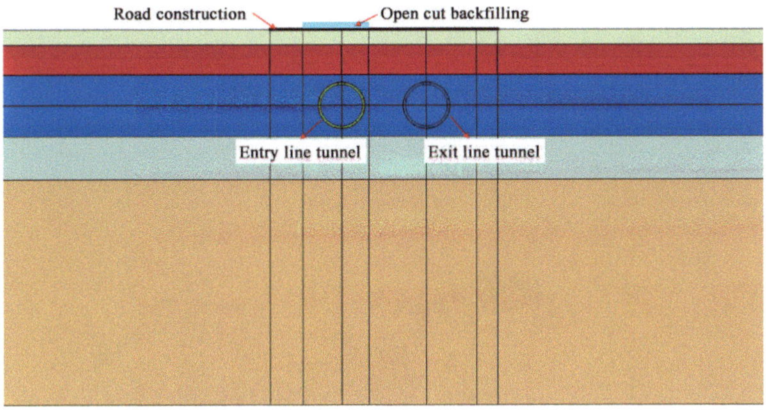

Fig. 5.34 Numerical calculation model

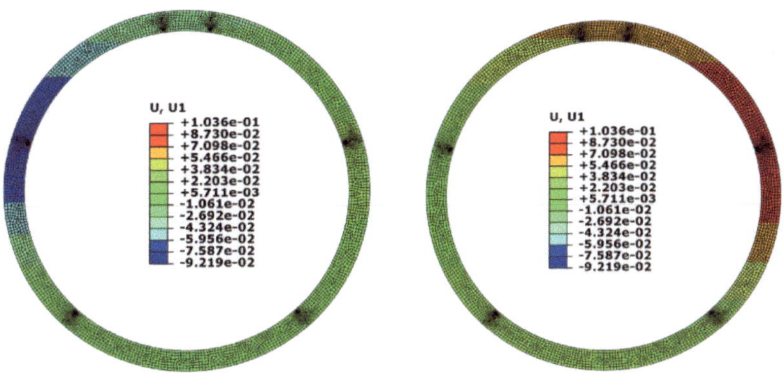

(a) Pedestrian Line Tunnel (b) Exit Line Tunnel

Fig. 5.35 Segment lateral deformation

5.3 Unloading Case and Control Remediation

5.3.1 Case 1: Zhongcao Road No. 106 Plot Development Project

(1) Project Overview

The Zhongcao Road No. 106 plot development project consists of a 15-story office tower and a high-rise podium building. The tower has 2 underground levels, while the podium building has 1 underground level. The total land area of the project is 8,907 m^2, as shown in Fig. 5.36. The total area of the foundation pit project

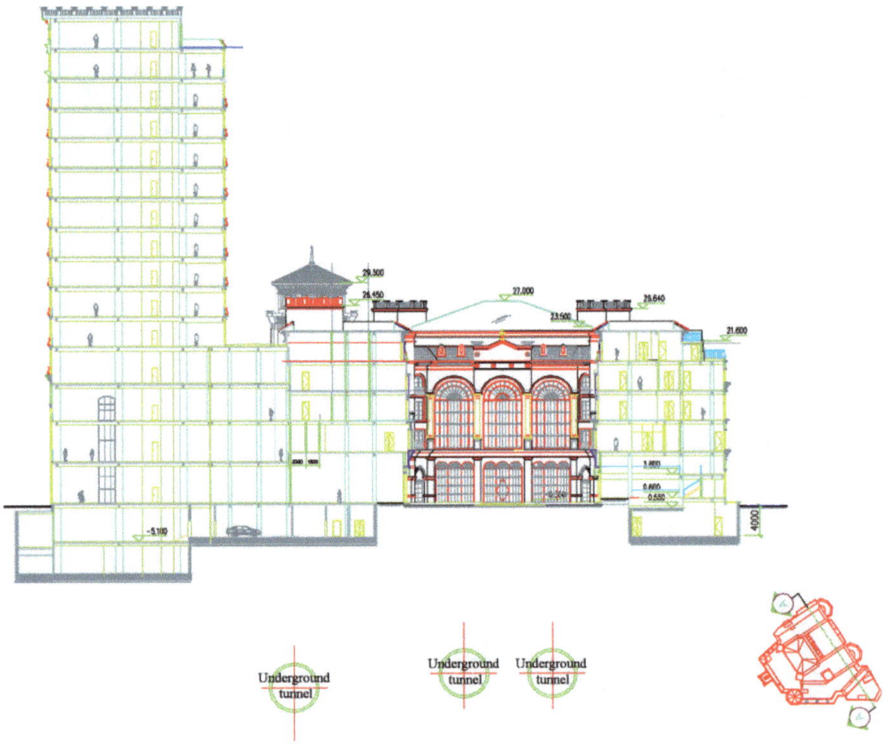

Fig. 5.36 Zhongcao road no. 106 plot development project

is approximately 5,500 m^2, with the tower's deep pit excavation area being about 950 m^2 and the excavation depth around 10.75 m. The podium building's shallow pit excavation area is about 4,550 m^2, with an excavation depth of approximately 5.30 m.

The project has three tunnels below the basement, including a section of the operational Metro Line 4 and the entrance and exit parking lot tunnels, as shown in Fig. 5.37.

The section tunnels beneath the foundation pit have a length of 50–60 m and a tunnel diameter of 6.2 m. The burial depth of the tunnel ceiling for the up and down lines is between 14.9 and 15.4 m, with the tunnel ceiling approximately 9 m below the basement floor. The parking lot line section tunnel has a burial depth of 17.7–19.5 m, and the tunnel ceiling is approximately 12.4 m below the basement floor.

(2) Construction Challenges and Corresponding Countermeasures

The site lacks soil layers ⑥ and ⑧, and the micro-confined water layer and confined water layer are connected. The micro-confined water layer needs to be depressurized. A vacuum deep well pump dewatering method is used for the deep pit, with a well drilled approximately every 150–200 m^2. Based on the excavation conditions, the groundwater level is lowered to 1.0 m below the excavation surface, and dewatering

Fig. 5.37 Plan relationship between foundation pit and metro line 4 section tunnels

is conducted for more than two weeks before the soil excavation begins. Short filter pipes are used for draining the micro-confined water to reduce the adverse impact of dewatering on the surrounding environment.

The project is divided into two parts: deep pit and shallow pit. First, the deep pit is excavated, and after the entire structure of the deep pit's basement is completed, the shallow pit is excavated. A partition wall separates the deep pit and shallow pit. Once the structure of the shallow pit's basement is completed, the partition wall will be removed in sections from top to bottom according to construction needs.

The shallow pit is located directly above the three sections of Metro Line 4 tunnels, with a complex relationship with the metro system. The soil within the shallow pit is fully reinforced, as shown in Fig. 5.38. Considering the overall construction sequence of the shallow pit, the size of each section, and the impact of excavation unloading on the metro tunnels, the pit is divided into five sections. Each section is further subdivided into smaller blocks, with timed excavation for each block and the bottom slab poured to control the tunnel's heave deformation. The foundation pit sections are shown in Fig. 5.39.

The foundation pit excavation for soft soils in the Shanghai area follows the "spatiotemporal effect" theory, strictly implementing block-based limited-time excavation and pouring support requirements. During excavation, support, and cushioning

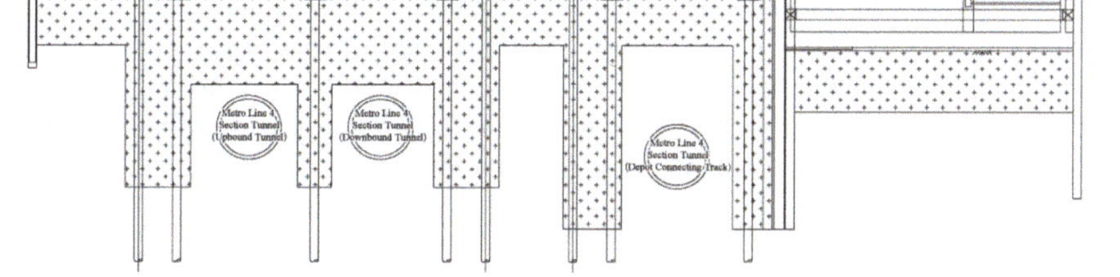

Fig. 5.38 Foundation pit reinforcement schematic

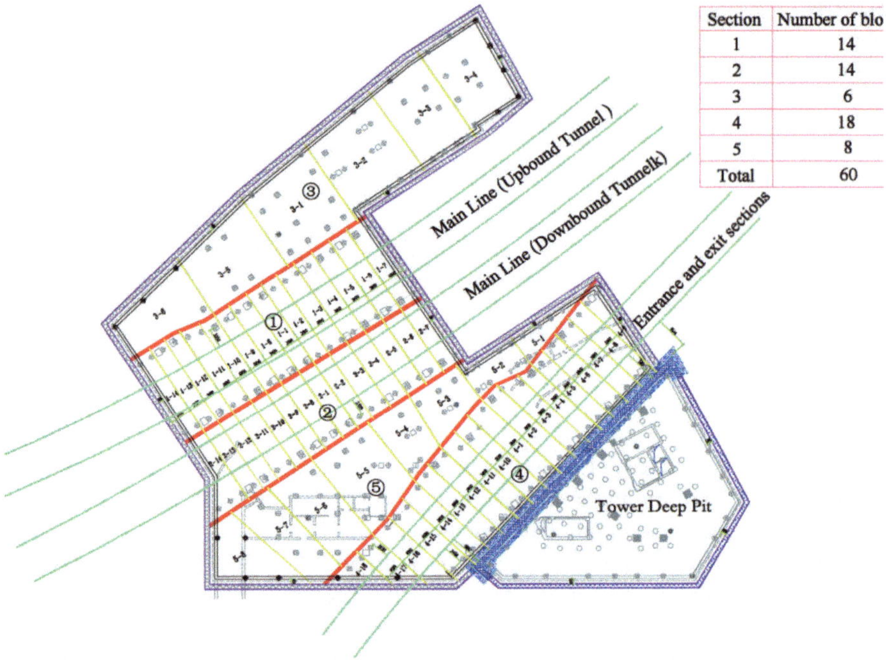

Fig. 5.39 Foundation pit section construction plan

layer construction, the general principles of "layer-by-layer, block-by-block, leaving soil for protection, symmetrical, limited-time excavation, and supporting excavation" are followed. The spatiotemporal effect principle is used to reduce the exposure time of unsupported foundation pit areas and strictly control the deformation of the pit.

The shallow pit is divided into five sections. Excavation begins in the middle of the pit. For the middle section, excavation is done one block at a time, followed by pouring a cushioning layer and bottom slab. After timely ballasting, the next block is excavated. The total ballasting time during excavation should be controlled within 12 h. Once the excavation of the middle section is complete, and the cushioning layer and bottom slab concrete are poured, steel pipe supports are set up in the surrounding soil at the design locations (one end placed on the already poured bottom slab corbel). Afterward, the remaining soil in the surrounding area is excavated in blocks, and the cushioning layer and bottom slab concrete are poured in time.

After the first-floor slab of the podium building is poured, one-third of the sandbag ballast is removed. After the second-floor slab of the podium building is poured, another third of the sandbag ballast is removed. The remaining sandbag ballast is removed after the third-floor slab structure is completed.

(3) Metro Monitoring and Analysis During Construction

The project was paused after the completion of pile foundation enclosure in 2012 and resumed in 2017. The construction schedule is shown in Table 5.6.

Table 5.6 construction working conditions overview for Zhongcao road no. 106 plot project

Date	Working conditions	Date	Working conditions
2017.11.23	Deep foundation pit excavation	2018.9.21	Shallow foundation pit Section 4 excavation
2017.12.31	Deep foundation pit bottom slab completed	2018.10.19	Shallow foundation pit Section 4 large bottom slab completed
2018.1.15	Deep foundation pit large bottom slab poured	2018.10.20	Shallow foundation pit Section 5 excavation
2018.4.22	Deep foundation pit to ± 0.0	2019.1.6	Shallow foundation pit Section 5 large bottom slab completed
2020.1.9	Deep foundation pit structure capping	2018.11.16	Shallow foundation pit Section 3 excavation
2018.7.10	Shallow foundation pit Section 1 excavation	2019.1.2	Shallow foundation pit Section 3 large bottom slab completed
2018.9.4	Shallow foundation pit Section 1 large bottom slab completed	2019.12.1	Shallow foundation pit t ± 0.0
2018.8.10	Shallow foundation pit Section 2 excavation	2020.11.16	Shallow foundation pit structure capping
2018.9.11	Shallow foundation pit Section 2 large bottom slab completed	2021.3.10	Continuation of observation completed

Figure 5.40 shows the vertical heave deformation of the metro tunnel after excavation of the shallow pit. According to the working condition statistics, from July 10 to September 4, 2018, excavation in area ① above the up line led to a maximum heave of approximately 4.1 mm on the up line; from August 10 to September 11, 2018, excavation in area ② above the down line resulted in a maximum heave of approximately 2.0 mm on the down line; from September 21 to October 19, 2018, excavation in area ④ above the entry and exit lines led to a maximum heave of approximately 1.0 mm on the entry and exit lines.

Figure 5.41 presents the time history curve of vertical heave deformation of the metro tunnel during shallow pit excavation. From the time history curve, it is evident that as the shallow foundation pit above the tunnel is excavated and unloaded, the tunnel undergoes continuous heave deformation.

The excavation depth of the podium building's shallow pit is 5.3 m. The ratio of tunnel heave amount (s) to excavation depth (h) is shown in Table 5.7. According to the results in Table 5.7, with the soil in the foundation pit above the tunnel reinforced using full reinforcement with mixing piles, and strict adherence to the time-limited excavation for each section, the heave deformation of the tunnel can be controlled to within 1.00‰ of the excavation depth.

Figure 5.42 shows the lateral convergence deformation of the metro tunnel after the excavation of the shallow pit. From the figure, it can be observed that the lateral convergence deformation decreased compared to before excavation, with a reduction range of −1.0 to −2.0 mm, indicating that the tunnel exhibits an inward shrinkage deformation trend along the transverse direction under the unloading effect directly

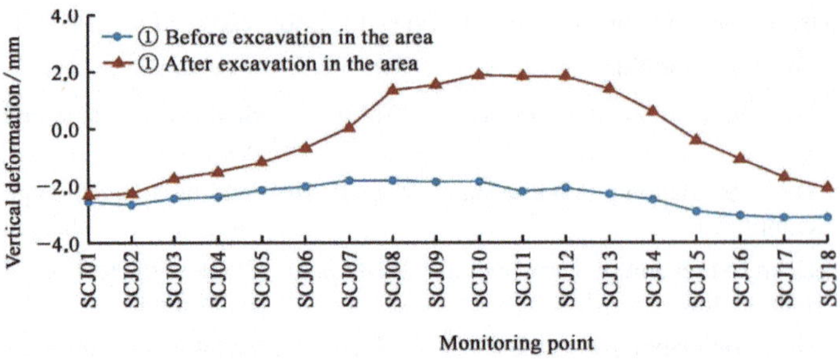

(a) Up line

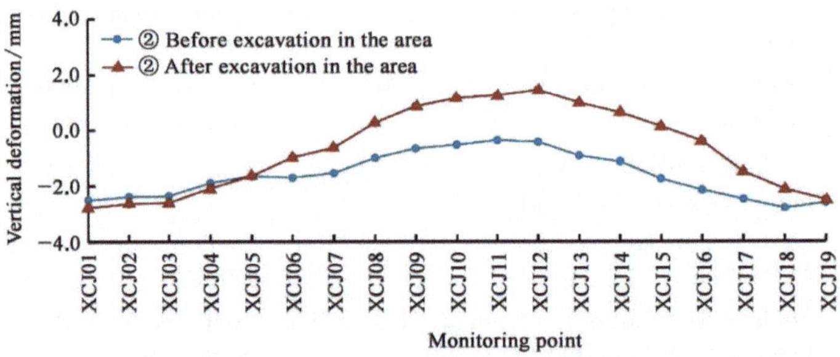

(b) Down line

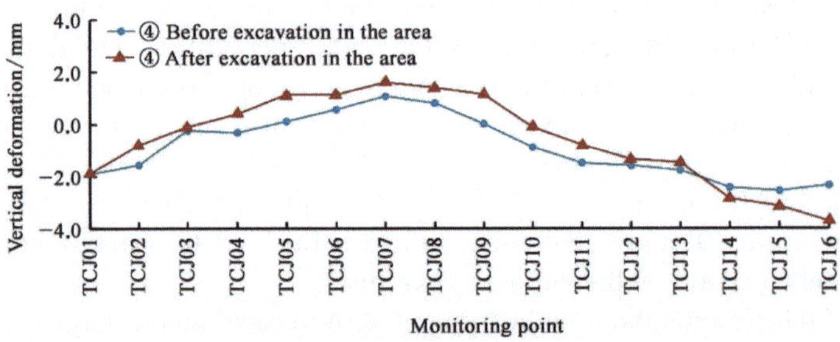

(c) Entry and exit line

Fig. 5.40 Vertical heave deformation of metro tunnel

above. Combining this with the deformation trend of the tunnel's bottom heave, it suggests that the tunnel segments show a "vertical duck egg" deformation pattern under the unloading effect directly above.

(4) Numerical Simulation Analysis

The numerical simulation method for shield tunnel deformation under unloading conditions described in Chap. 4 is used to establish a finite element model, as shown in Fig. 5.43. During the calculation process, the construction conditions are simulated, meaning that the deep foundation pit is excavated first, followed by the excavation of the shallow foundation pit in sections. Before excavation, full reinforcement of the

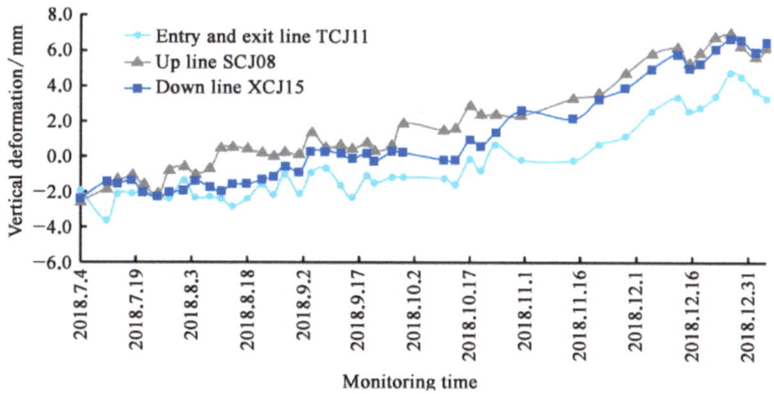

Fig. 5.41 Time history curve of vertical heave deformation of metro tunnel

Table 5.7 Tunnel heave amount ratio

	Up line	Down line	Entry and exit line
Maximum vertical heave amount *s*/mm	+4.1	+2.0	+1.0
(s/h)/‰	0.77	0.38	0.20

shallow pit soil and reinforcement of the soil on both sides of the tunnel are considered. The HSS model is used for the soil layers, with model parameters determined based on the project's geological survey report. Tunnel segments are modeled using a beam-hinge joint model.

Figure 5.44 shows the deformation results of the tunnel segments caused by the excavation of the foundation pit directly above. The data in the figure represent the deformation amounts at the top, bottom, and waist of the tunnel, with the positive and negative signs indicating direction. The positive direction is defined as the normal direction of the segment ring, with positive values indicating displacement towards the interior of the segment and negative values indicating displacement towards the exterior normal direction of the segment.

From the figure, it can be observed that during the excavation stage of the foundation pit above, the heave on the up line is approximately 2.0 mm, the heave on the down line is approximately 2.2 mm, and the heave on the entry and exit lines is approximately 2.3 mm. The calculated heave on the up line is slightly smaller than the measured maximum heave. Based on the working condition records, it is noted that there was cross-construction in the shallow pit Section ① and Section ②, which affected the up line, an aspect not considered in the numerical simulation analysis.

According to the calculation results, when the unloading ratio is between 0.3 and 0.35, the calculated heave-to-depth ratio is approximately 0.4‰.

The relative deformation at the waist of the tunnel is the lateral convergence deformation of the tunnel. Based on Fig. 5.44, the lateral convergence deformation caused by unloading above the tunnel is shown in Table 5.8, with negative values indicating lateral inward shrinkage, with the shrinkage amount being approximately

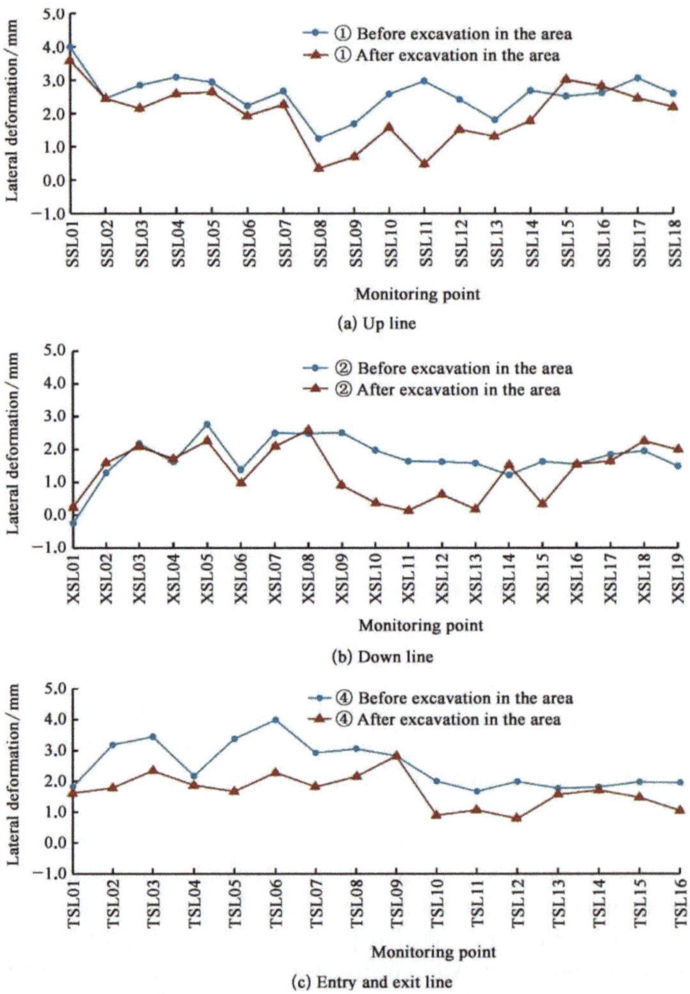

Fig. 5.42 Lateral convergence deformation of metro tunnel

−2.0 mm. Under the unloading effect above, the tunnel experiences overall vertical stretching and lateral inward shrinkage, presenting a "vertical duck egg" deformation pattern.

The case studies show that when the tunnel burial depth is large and the unloading ratio is small $\left(n < \frac{1}{3}\ \frac{1}{2}\right)$, measures such as sectioning, reinforcement, piling, and time-limited excavation to form the bottom slab can effectively control the tunnel heave amount to less than 1‰.

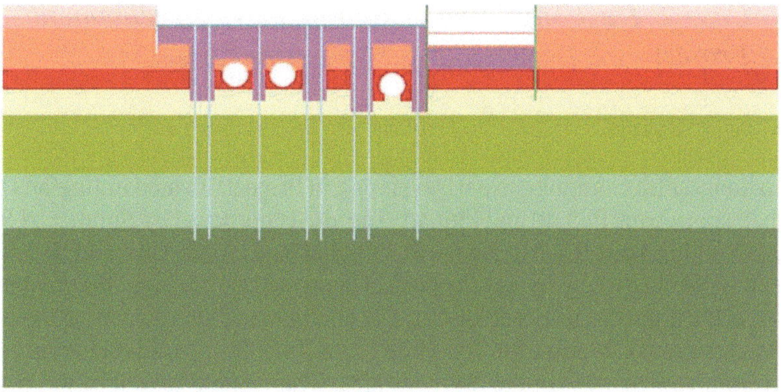

Fig. 5.43 Finite element calculation model for the Zhongcao road project

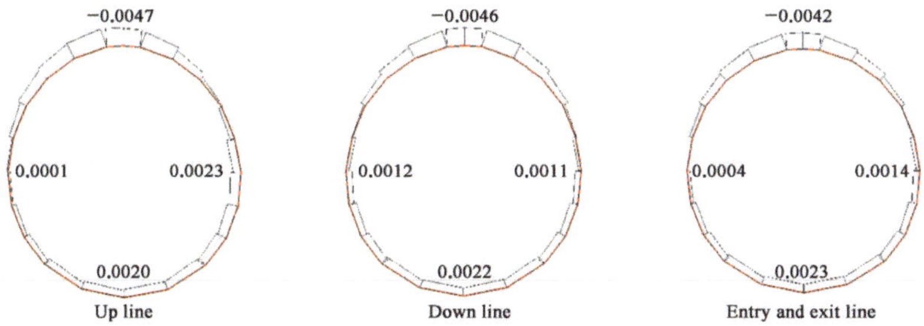

(a) Shallow Pit Section ① Excavation (b) Shallow Pit Section ② Excavation (c) Shallow Pit Section ④ Excavation

Fig. 5.44 Tunnel segment deformation caused by unloading above (unit: m)

Table 5.8 Tunnel deformation statistics for Zhongcao road project

Line	Unloading ratio n	Vertical heave amount s/mm		Lateral convergence Δd/mm		Heave-to-depth ratio/‰	
		Measured	Calculated	Measured	Calculated	Measured	Calculated
Up line	0.35	+4.1	+2.0	−2.5	−2.4	0.77	0.38
Down line	0.35	+2.0	+2.2	−1.6	−2.3	0.38	0.42
Entry and exit line	0.30	+1.0	+2.3	−1.7	−1.8	0.20	0.43

5.3.2 Case 2: Dazhongli Plot Comprehensive Development Project

(1) Project Overview

The Dazhongli Plot Comprehensive Development Project, located in Jing'an District, Shanghai, spans over Plot 40 and Plot 46. The site is bordered by Shimen Yi Road to the west, Nanjing West Road to the north, Qinghai Road to the east, and Weihai Road to the south. Wujing Road divides the project area into two sections: the southern area (Plot 40) and the northern area (Plot 46). The southern area is adjacent to the Nanjing West Road station of Metro Line 13, while Metro Line 2's operational tunnel crosses the site from east to west in the northern area. The general layout of the project is shown in Fig. 5.45. The tunnel diameter for this section is 6.2 m, with a wall thickness of 350 mm, and the tunnel's top burial depth is approximately 9.0 m.

The South Area Plot 40 (Nanjing East Road Station on Line 13), North Area Plot 46, and the shield tunnel of Line 13 (underpassing Line 2) will have a significant impact on Metro Line 2 during construction, with the greatest risk posed by Plot 46. This case study analyzes the excavation impact in the northern area (Plot 46) of the project. As shown in Fig. 5.46, the site area is approximately 4,300 m^2, and it is planned to construct a one-story and four-story leisure and commercial plaza. The one-story building falls within the B-block area of the project, using a natural foundation, with an excavation area of approximately 350 m^2, and the bottom slab burial depth is about 1.55 m. The excavation depth for Area A is between 3.95 and 4.55 m. Since the building is located above the operational Metro Line 2 tunnel, a pile foundation is used for the four-story building, and the foundation slab is approximately 4.45 m above the top of the operational tunnel. The section tunnel below the foundation pit has a length of about 102 m, and the excavation area is approximately 3,810 m^2.

(2) Construction Challenges and Corresponding Countermeasures

The retaining piles use Φ850 SMW method piles with internal insertion of 700 × 300 H-shaped steel (inserted every other pile), with pile depths of 7.5 and 11.5 m. The reinforcing range of the portal frame in the middle and both sides of the tunnel extend to 17.8 m below the ground (3 m below the tunnel bottom), with the nearest reinforcement being 1.5 m from the tunnel. A total of 143 Φ850 mm bored piles are used, with an effective pile length of 71.1 m, and the pile ends are treated with post-grouting. These piles are distributed on both sides of the tunnel. The remaining area is fully reinforced to a depth of 7.5 m, with the reinforcement base 1.1 m below the tunnel top. The relationship between the pile foundation, soil reinforcement range, and tunnel profile is shown in Fig. 5.47.

The project construction site is narrow, with difficulties in material turnover and conflicts between the on-site conditions and the designed construction process. The foundation slab of Area A must be excavated and completed before the foundation slab of Area B can be constructed. To effectively control the impact of the excavation

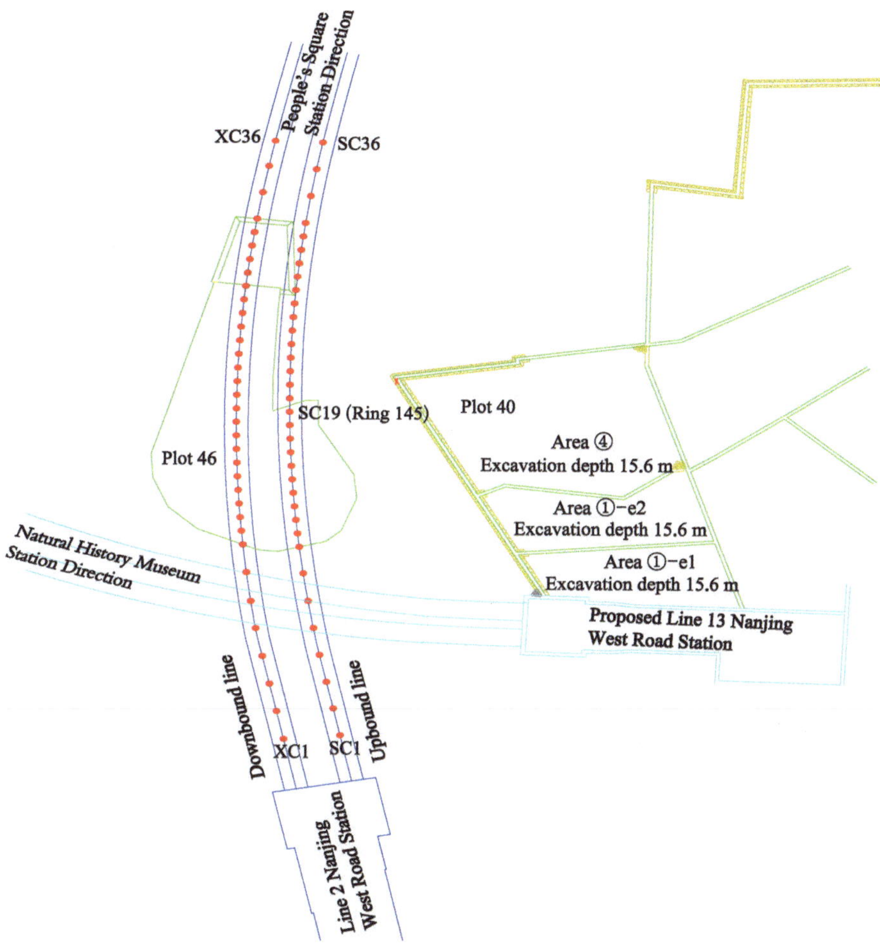

Fig. 5.45 General layout of project location

on Metro Line 2 in Plot 46, the excavation work is divided into 59 small sections, each controlled within 60 m^2. Excavation is carried out by first constructing the sections above the down-line tunnel, followed by the sections above the up-line tunnel, with the work process interspersing sections outside the tunnel area. Each section has a width of about 3 m, ensuring that at least one pile is located at each end of the section. Excavation for each section must not exceed 7 h, and the work must be completed during the metro's non-operational period. The partitioning of the Area A foundation pit is shown in Fig. 5.48.

To ensure the time nodes and deformation control requirements for each construction section, the technical plan has been refined as follows:

① Optimized the reinforcement joint lap scheme and consolidated the reinforcement specifications.

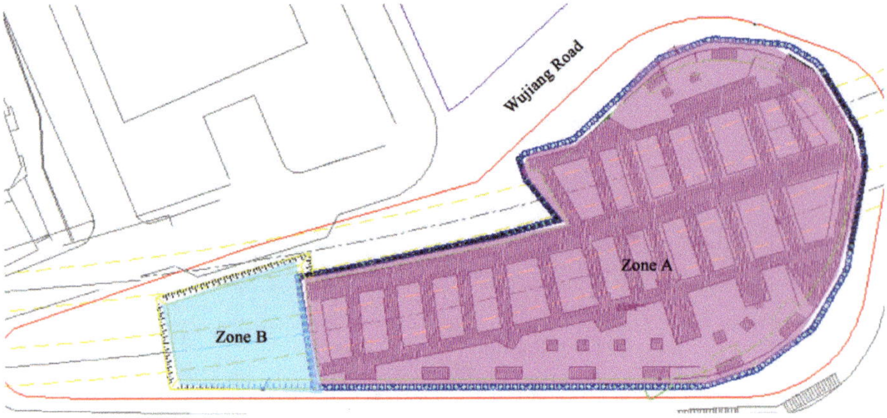

Fig. 5.46 Zoning plan of plot 46# foundation pit excavation

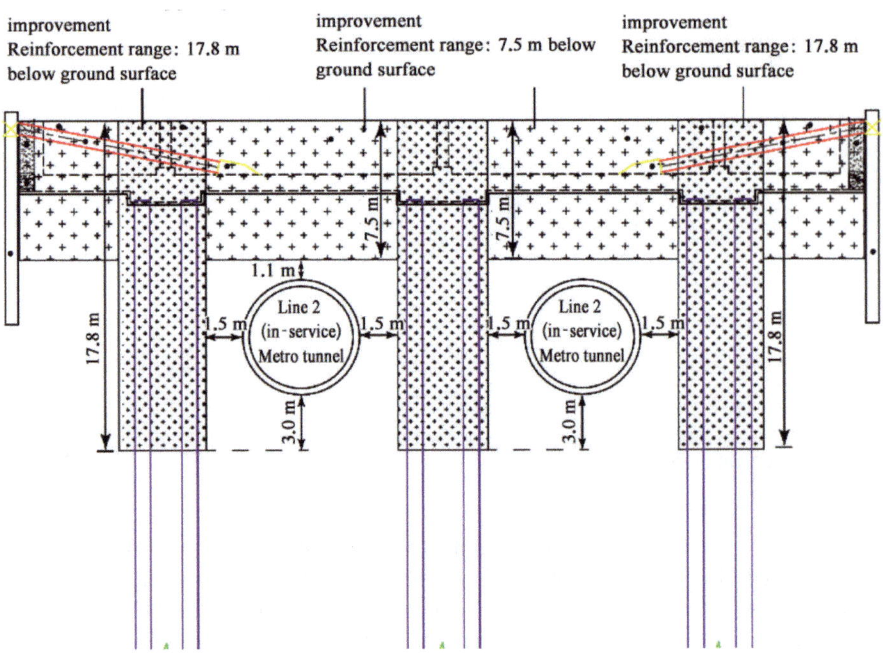

Fig. 5.47 Sectional relationship between plot 46# foundation pit and line 2 metro tunnel

② Adopted a flat slab foundation.
③ Used 10.0 cm PS board and 2.0 cm nine-layer template for the cushion layer.
④ For waterproofing, expanded water-stop strips are placed in each section.

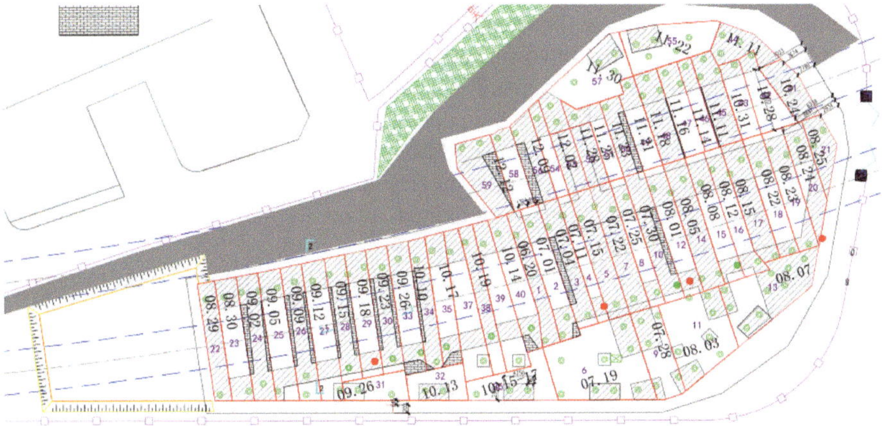

Fig. 5.48 Zoning areas and construction timeline

⑤ During the actual first construction, the installation of embedded parts significantly impacted the working hours. The use of positioning plates to group and fix embedded bolts effectively saved time.

Throughout the process, the construction measures have also been optimized:

① Earth excavation: The soil has no lateral confinement and reaches a compressive strength of 2.5 MPa, making excavation challenging. A combination of pick machines, backhoe excavators, and manual labor was used. Before formal excavation of each section, the topsoil above the area was pre-drilled and crushed during the day to speed up nighttime excavation.
② Base slab reinforcement prefabrication and binding: Once the cushion layer was laid, it was immediately hoisted and placed.
③ Pile drilling and concrete pouring work: Sufficient workers were assigned to manually chisel according to excavation progress and construction operations. Concrete was prepared in advance and poured promptly.

(3) Metro Monitoring and Analysis During Construction

After the drilling and grouting pile construction and the reinforcement of the protective slurry piles, section-by-section construction began on June 20, 2013. The down-line tunnel was completed on October 19, 2013, and construction on the up-line sections above started. The foundation pit excavation was completed on December 11, 2013, and the overall base slab was formed, taking six months.

For the excavation work in Plot 46, based on data curve analysis for the time nodes of the excavation sections and lines, it was observed that the soil above the Line 2 tunnel was lost during excavation. The reduction in the overburden weight caused neighboring soil to rebound. Specifically, the soil above the down-line tunnel between June 20 and October 19 had a cumulative heave of up to 4.0 mm, while the change in the up-line was small, less than 1.0 mm. Between October 19 and

December 11, the soil above the up-line tunnel was excavated and had a cumulative heave of 3.0 mm, while the down-line showed a smaller continuous heave of about 1.0 mm. Overall, the project remained in a safe and controllable state.

During the early stages of the project, due to the piling, reinforcement, and protective construction, the tunneling already showed heave effects, and the down-line tunnel had a significant change, with a maximum value of +12.2 mm. The convergence of tunnel diameter changes remained relatively stable, with a cumulative maximum value of +7.0 mm.

Figures 5.49 and 5.50 show the automatic horizontal settlement monitoring data for the up-line and down-line.

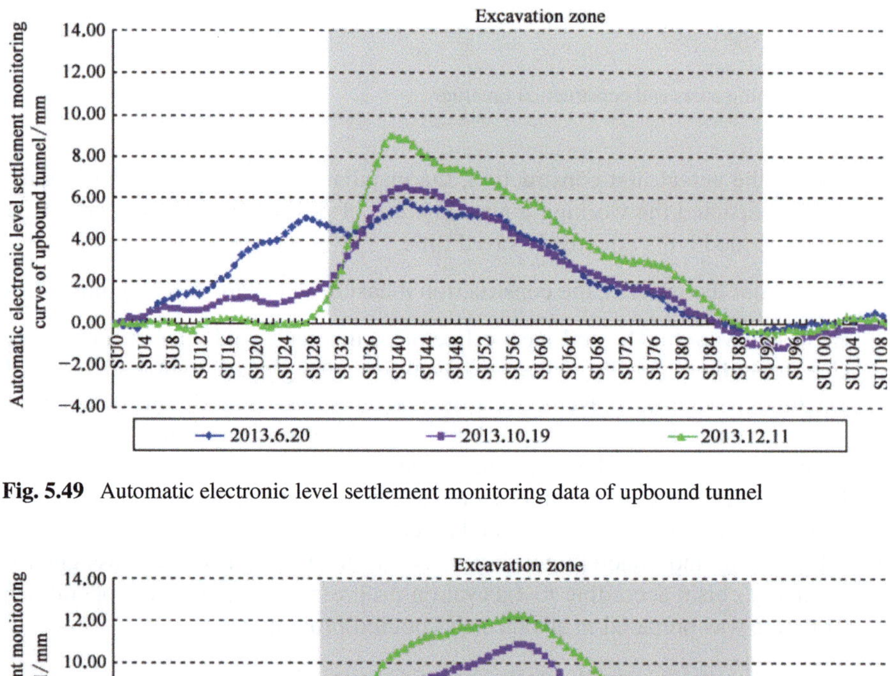

Fig. 5.49 Automatic electronic level settlement monitoring data of upbound tunnel

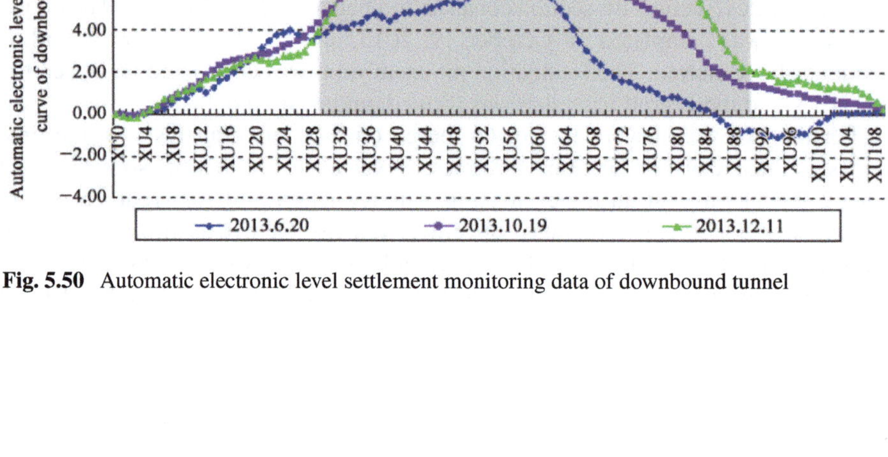

Fig. 5.50 Automatic electronic level settlement monitoring data of downbound tunnel

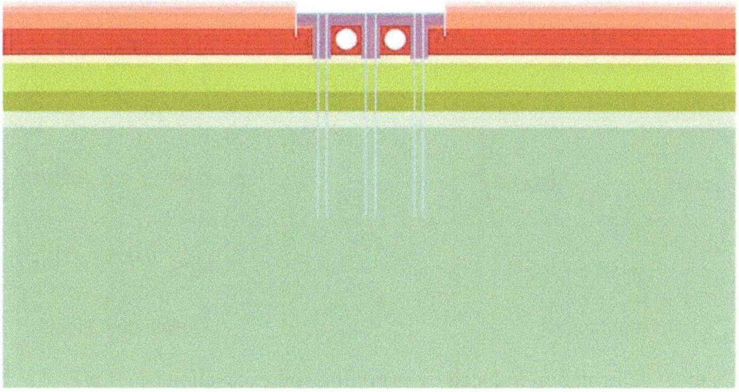

Fig. 5.51 Finite element calculation model of Dazhongli project

(4) Numerical Simulation Analysis

The numerical simulation method for shield tunnel deformation under unloading conditions, as introduced in Chap. 4, is used to establish a finite element numerical simulation model, as shown in Fig. 5.51. The model takes into account soil layering, and during the calculation process, construction conditions are simulated. Specifically, the soil above the down-line tunnel is excavated first, followed by the excavation of the soil above the up-line tunnel, with excavation on both sides of the tunnel also interspersed. The construction conditions also account for full reinforcement of the soil within the pit and anti-pull piles on both sides of the tunnel. The soil layers are modeled using the HSS model, with model parameters determined based on the project's geological survey report. Tunnel segments are modeled using a beam-hinge joint model.

Figures 5.52 and 5.53 show the tunnel segment deformation caused by excavation of the soil above the down-line and up-line tunnels, respectively. From the figures, it can be observed that when the soil above the down-line tunnel is excavated, the tunnel segments of the down-line show a vertical stretching deformation trend, while the deformation trend of the neighboring up-line tunnel tends to exhibit rigid body displacement towards the foundation pit side. Similarly, when the soil above the up-line tunnel is excavated, the tunnel segments exhibit the same deformation trend.

Table 5.9 lists the tunnel segment deformation caused by the excavation of the soil directly above the tunnel in the Dazhongli project. According to the calculation results, when the unloading ratio is 0.5, the tunnel heave amount is approximately 1.3‰ of the excavation depth.

It is important to note that appropriate measures such as reinforcement, piling, sectioning, and time-limited excavation are highly effective in controlling tunnel hazards. However, foundation reinforcement is also a form of disturbance that exerts compressive forces on the tunnel, requiring optimization during the construction process.

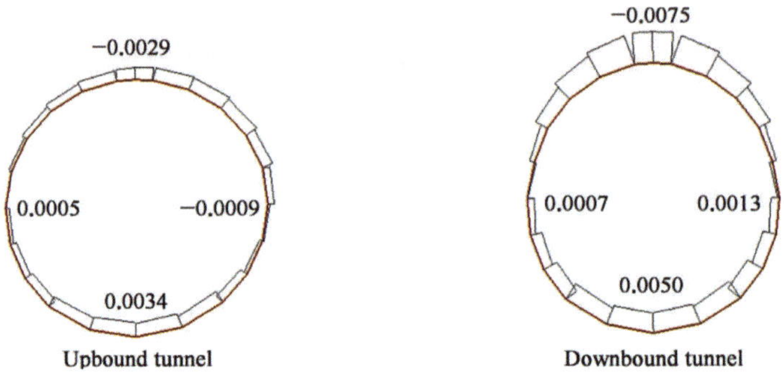

Fig. 5.52 Segment deformation induced by overlying soil excavation (unit: m)

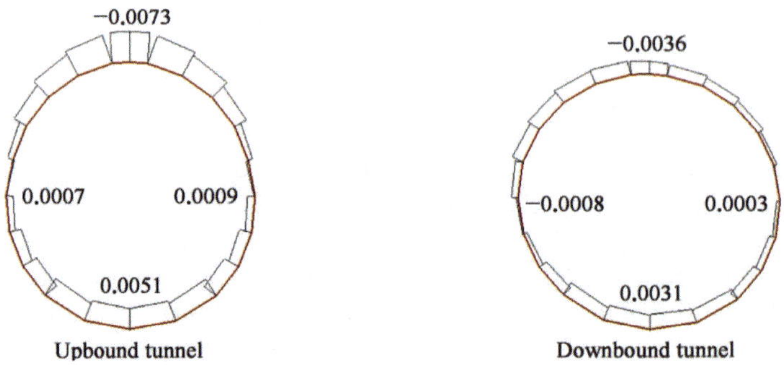

Fig. 5.53 Segment deformation induced by overlying soil excavation (unit: m)

Table 5.9 Tunnel deformation statistics of Dazhongli project

Working conditions	Unloading ratio n	Vertical uplift s/mm		Horizontal convergence Δd/mm		Uplift-depth ratio/‰
		Upbound tunnel	Downbound tunnel	Upbound tunnel	Downbound tunnel	
Soil excavation above downbound tunnel	0.47	+3.4	+5.0	+0.4	−2.0	1.25
Soil excavation above upbound tunnel	0.47	+5.1	+3.1	−1.6	+0.5	1.28

5.3.3 *Case 3: Dongfang Road Underpass Project*

(1) Project Overview

The Dongfang Road Underpass Project is located at the intersection of multiple important roads, including Century Avenue, Zhangyang Road, and Dongfang Road in Pudong, as shown in Fig. 5.54. The project runs approximately in a north–south direction with a total length of 600 m. The intersection with Metro Line 2 is located around 100 m west of Dongfang Road Station, crossing diagonally over the tunnel. The underpass is divided into two structural forms: a rectangular box-type structure and an open dock-type structure. The box-type structure is 200 m long, located at the intersection, and the north and south ends are open dock-type structures, with a length of 270 m. The underpass is divided into 23 sections, S01 to S12 and N01 to N11, with an excavation depth of 6.3 m. The underpass is designed with a bi-directional, 4-lane road, and a bi-directional, 4-lane road surface is constructed on both sides of the tunnel. To ensure the passage of utilities beneath Dongfang Road, two east–west utility tunnels are constructed on the north and south sides of the project. The northern tunnel has dimensions of 95.2 m × 22.0 m (length × width), with a foundation bottom elevation of −5.90 m, and the southern tunnel has dimensions of 80.2 m × 15.3 m (length × width), with a foundation bottom elevation of −6.30 m at its lowest point. Due to the large number of utilities passing through the entire project construction area, with underground structures such as the operational Metro Line 2 and the under-construction Line 4, as well as the reserved station for Line 9, the excavation of Sections N01 (south side) and N02 (north side) passes directly over Metro Line 2, with the base slab of the structure being less than 2.6 m from the top of the Metro Line 2 tunnel. Due to the significant unloading effect from the foundation pit excavation, the rebound of the soil layers at the base of the pit drives the metro tunnel upwards. The excavation construction poses a significant risk to the operational safety of Metro Line 2, and even a small mistake could potentially endanger the safety of the metro. Therefore, controlling the tunnel settlement and heave caused by unloading during construction is the key to the design and construction.

(2) Construction Challenges and Corresponding Countermeasures

Due to the excavation depth of 6.3 m, the overburden above the tunnel is only 2.6 m thick. The unloading effect from the excavation exceeds 11 tons, and large-depth excavation unloading directly causes the soil layers below the foundation pit to rebound, affecting the metro tunnel. According to calculations, without effective measures to control the deformation of the tunnel, the heave will exceed the tunnel deformation control limit, and the tunnel structure's deformation will not meet the operational safety requirements of the metro. This construction challenge must be addressed from both design and construction perspectives. The construction involving the operational metro tunnel, especially the tunnel section directly above Metro Line 2 and the deep foundation pit construction of the utility tunnel on the north side of Zhangyang Road, has the most significant impact on the metro tunnel. This section's construction is the key to the success of the project.

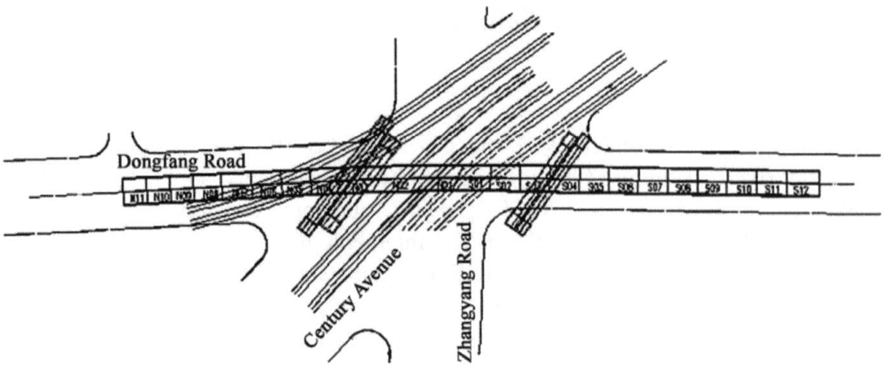

Fig. 5.54 Planar relationship between underpass and tunnel

The foundation pit retaining structure that crosses over Metro Line 2 uses the SMW method piles, which offer high construction efficiency, good waterproofing, and relatively low cost. However, in the section crossing over the metro tunnel, the pile insertion rate is insufficient to meet the foundation pit's anti-sliding safety. Therefore, the soil in the 10-m zone outside the foundation pit retaining structure must be reinforced to ensure the pit's safety. To reduce and effectively control the tunnel heave, the soil within the foundation pit crossing over the metro tunnel must be fully reinforced. The reinforcement uses jet grouting piles and SMW method piles, with a cement content of no less than 20%. To ensure the quality of the reinforcement, quality samples must be taken before excavation. The distance between the reinforcement and the tunnel top, as well as the tunnel sides, must be strictly controlled within 0.5–0.7 m. The foundation reinforcement profile is shown in Fig. 5.55.

To effectively control the tunnel heave during excavation, H-shaped steel is inserted into the SMW piles along both sides of the tunnel, forming an effective constraint system with the reinforced soil, to control the tunnel heave during soil excavation. The depth of the steel insertion must not be less than 25 m. The soil excavation through the tunnel section is carried out separately for the up and down lines, dividing the soil into several sections along the longitudinal direction of the tunnel for excavation. The excavation is divided into two layers: the first layer has an excavation depth of no more than 2 m, and the remaining soil is the second layer, to control the early unloading. The excavation process follows strict principles of "layered, sectioned, symmetrical, balanced, and time-limited" excavation, supporting, and back-pressing.

During excavation, the middle partition wall is excavated first to form the partition wall quickly, and then excavation proceeds in two directions along the longitudinal tunnel. The excavation width of each section is generally no more than 3–4 m. For the second layer of excavation, the excavation time for each section is controlled to 3 h, with no cushion layer used to control the total construction time for excavation and concrete pouring. The bottom slab reinforcement is prepared in advance before excavation, and concrete with a higher grade and slight expansion properties is used.

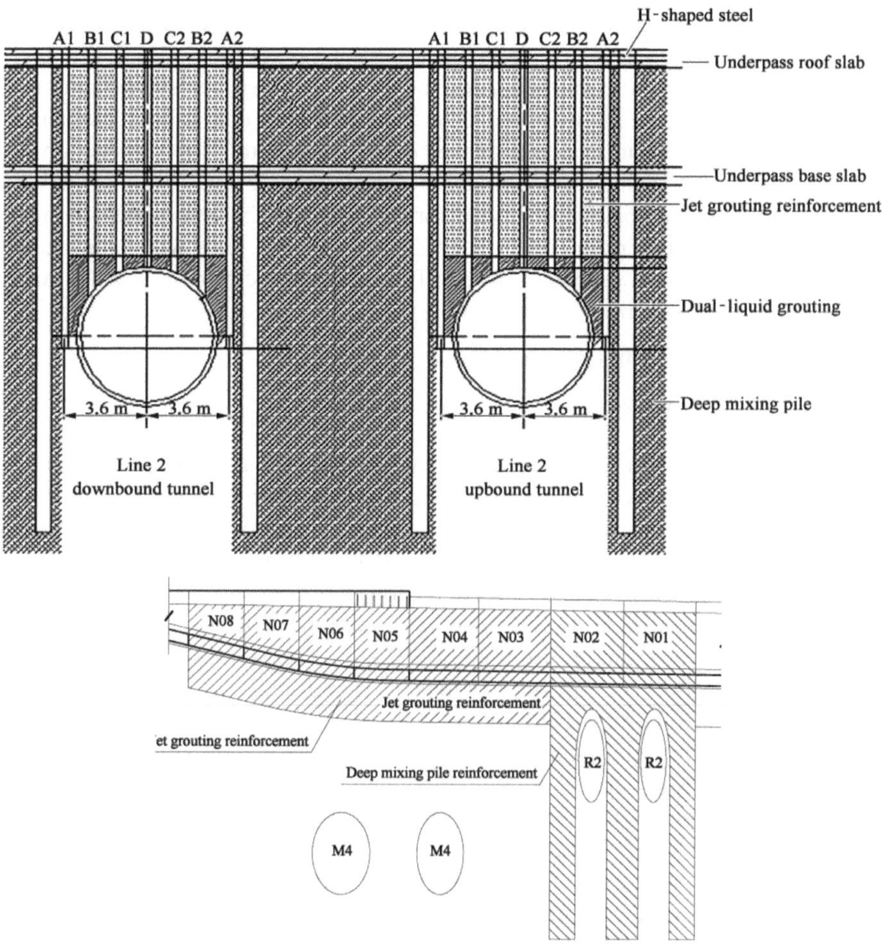

Fig. 5.55 Ground improvement profile

After concrete pouring, sandbags are used for back-pressing, and the total construction time for each section is controlled to 7–9 h. The construction requirements between adjacent sections are the same as previously mentioned.

Due to the very close proximity of the construction to the metro tunnel, even a small mistake could potentially endanger metro safety. Each step of the construction is highly risky. During the implementation of jet grouting and mixing piles, the position and verticality of the piles must be accurately located, and strict control must be maintained over the pile depth. The tunnel position and the overburden depth must be verified on-site by the metro operating unit. Construction above and near the tunnel must only be performed after the trains have stopped running. Preliminary tests must be conducted to assess the impact on the tunnel, and it must be decided whether to proceed with construction during the day or at night based on the results. The construction sequence must also be optimized. Additionally, real-time monitoring is required during construction, with immediate alerts in case of anomalies. The tunnel's diameter convergence is monitored using the Basset convergence system, and real-time high-precision leveling instruments are used to monitor the tunnel's settlement and heave, with deformation data displayed every 10–20 min. The construction unit is required to further refine the construction organization design, organize the work carefully, and prepare contingency plans for any situation that may endanger metro operation, ensuring that each construction step is operable. This will minimize the foundation pit exposure time and effectively control the deformation of the metro structure, ensuring safe metro operation.

(3) Metro Monitoring and Analysis During Construction

(1) Foundation Pit Retaining Structure and Foundation Reinforcement

On the night of April 25, 2003, experimental grouting of the down-line tunnel roof was conducted using jet grouting piles, and real-time monitoring was performed. During the grouting pile construction, the tunnel structure's roof experienced an instantaneous maximum downward displacement of 2.5 mm, and the displacement in the 45-degree direction on both sides of the top was between 1.5 and 1.7 mm. After stabilization, the maximum displacement was 1.0 mm, with lateral deformation of the tunnel at the waist ranging from 0.5 to 0.7 mm. The grouting pile construction significantly impacted the operational tunnel. Therefore, it was required that the grouting pile construction should only be carried out when the trains had stopped running, with optimization and adjustment of the construction parameters to meet the reinforcement requirements while minimizing the tunnel's deformation impact.

Meanwhile, experimental SMW pile construction was conducted on both sides of the tunnel cross-section, using Φ650 mm and Φ850 mm mixing piles, with the center of the pile 1.4 m away from the tunnel's edge. The pile length was 25 m. A single SMW pile caused compressive forces on the tunnel, with the side of the tunnel closer to the mixing pile experiencing a compression of 2.1 mm, while the far side experienced a compression of 0.4 mm, and the tunnel roof experienced an upward heave of 3.0 mm. Given the significant impact of the mixing pile construction on the tunnel, it was required that the first row of piles closest to the tunnel be constructed after the trains stopped running, and the piles should be constructed alternately. The other rows of mixing piles were constructed in the "near first, far later" sequence. Throughout the entire foundation pit reinforcement construction, deformation of the metro tunnel was continuously monitored using a static level gauge and the Basset system. The foundation pit reinforcement continued until June 30, and by the time the excavation of the soil above the tunnel started, the localized heave of the up-line and down-line tunnels had reached a maximum of 6–7 mm.

(2) Excavation of the Foundation Pit Above the Tunnel

Due to the large number of complex underground pipelines and their deep burial, as well as the project being located at a busy traffic intersection, the construction posed considerable challenges. To ensure timely completion, adjustments were made not

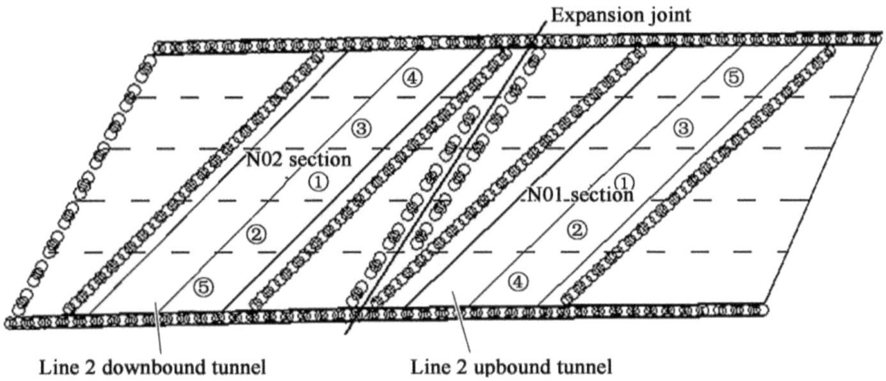

Fig. 5.56 Block construction of upbound and downbound tunnels

only to the construction sequence for the up and down lines but also to the original excavation width of 3–4 m. After thorough discussions, the minimum excavation width was appropriately expanded to shorten the construction period. Excavation of Section N01 was carried out first, followed by Section N02. The adjusted construction sequence for the up and down lines is shown in Fig. 5.56.

Starting from April 27, 2003, when the soil reinforcement was conducted, and continuing until the excavation began on July 15, the cumulative settlement of the up-line tunnel in some localized areas reached 7 mm. On July 15, 2003, construction began on the first layer of soil at the middle partition wall above the up-line tunnel, causing a rebound of 1.5–2 mm in the up-line tunnel. On July 17, excavation of the soil at the second layer of the middle partition wall started. First, the soil on both sides of the tunnel was excavated, followed by excavation of the soil directly above the tunnel. By 22:30, the excavation was completed, and the steel reinforcement and vertical formwork were tied. At 3:00 on the 18th, concrete was poured for the middle partition wall, completing at 5:00. By 9:00, the sandbags were back-pressed to a height of 3 m. During the entire construction process, the tunnel experienced continuous heave of 1–3 mm, with the cumulative heave amount reaching 12.1 mm. On the 20th, excavation of the second section of soil began, followed by the excavation of the third section on the 21st, and the fourth section on the 24th. By this time, the maximum heave had reached 16 mm. On the 25th, excavation of the fifth section of soil began. The pouring was completed by 6:00, and the back-pressing finished by 11:00, with the maximum cumulative heave reaching 17 mm. After the concrete for the middle partition wall and side walls was completed on August 2, the up-line tunnel remained relatively stable at 16–17 mm.

The time-history curve of the up-line tunnel heave from the foundation pit reinforcement to excavation and backfilling is shown in Figs. 5.57 and 5.58. The points labeled sCJ15-18 in the figures represent monitoring points located below the middle of the foundation pit. During this period, the down-line tunnel, due to the unloading effect from the up-line tunnel, experienced a heave of 6–7 mm before excavation, with the cumulative heave reaching 10 mm, and the effect reached more than 3.5 mm, as shown in Fig. 5.59.

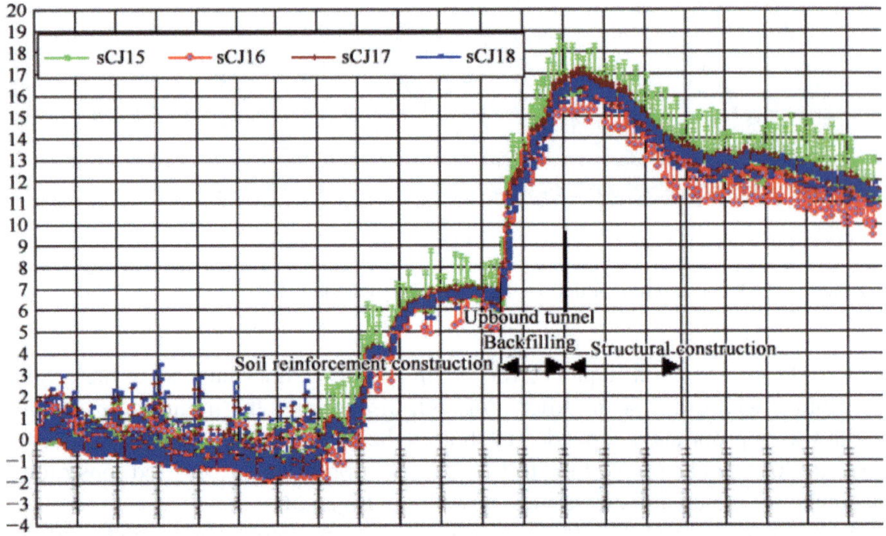

Fig. 5.57 Uplift time-history curve of line 2 upbound tunnel during foundation pit reinforcement-excavation-backfilling

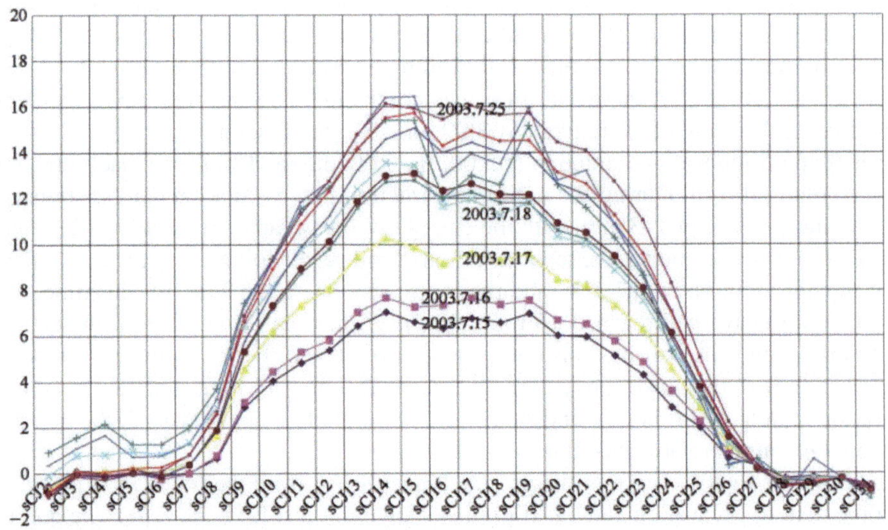

Fig. 5.58 Uplift curve of line 2 upbound tunnel during foundation pit reinforcement-excavation-backfilling

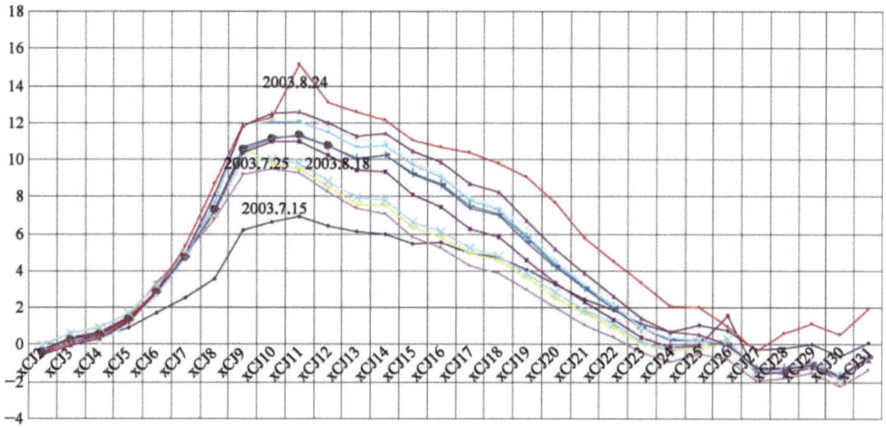

Fig. 5.59 Influence of construction factors on downbound tunnel deformation

Starting in late April 2003, reinforcement was carried out on the soil above and on both sides of the down-line tunnel of Metro Line 2. On August 18, excavation of the topsoil layer to a depth of 1.5–2.0 m was performed. On August 19, excavation of the first section of soil directly above the tunnel began at 19:00, and excavation was completed by 24:00. Concrete pouring for the section was completed at 4:00 on the 20th, and back-pressing finished by 10:00. Subsequently, one section of the base slab was constructed each day until the last section was completed on August 24. The construction requirements were the same as those for the up-line tunnel, and the concrete pouring for the top slab was completed by September 4. During this period, the heave of the up-line tunnel increased by 6 mm, bringing the cumulative heave to 15.24 mm.

The time-history curve of the down-line tunnel heave from the foundation pit reinforcement to excavation and backfilling is shown in Fig. 5.60. The points labeled xCJ12-xCJ17 in the figure represent monitoring points located below the middle of the foundation pit. From the figure, it is clearly shown that the excavation of the foundation pit above the up-line tunnel caused a heave of 3 mm in the down-line tunnel.

(4) Numerical Simulation Analysis

The numerical simulation method for shield tunnel deformation under unloading conditions, as introduced in Chap. 4, is used to establish a finite element numerical simulation model, as shown in Fig. 5.61. Since the Dongfang Road underpass project intersects Metro Line 2 at an angle of approximately 45°, a profile perpendicular to the tunnel alignment is selected for modeling. The model takes into account soil layering, full reinforcement of the soil within the pit before excavation, and grouting reinforcement of the soil within 1 m of the upper half of the tunnel. Additionally, anti-pull piles on both sides of the tunnel are considered based on the actual engineering conditions. During the calculation process, the model simulates segmented

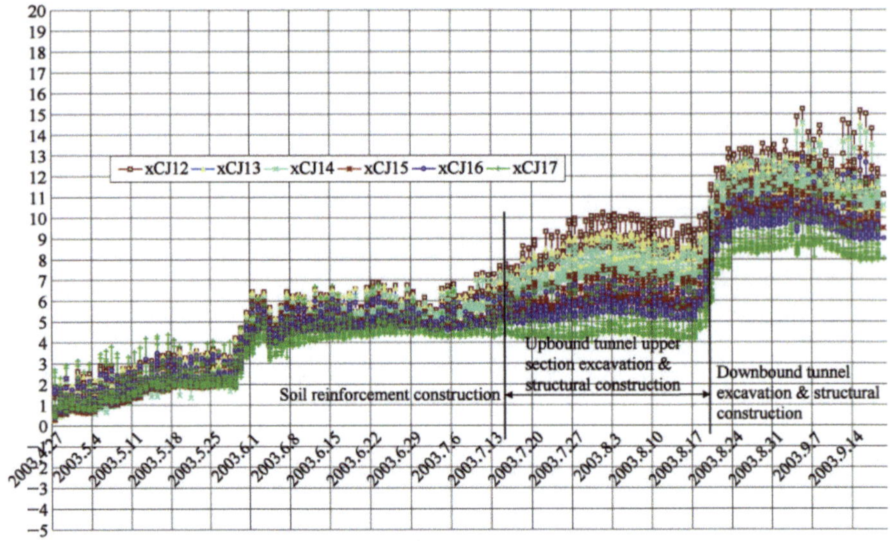

Fig. 5.60 Time-history uplift curve of line 2 downbound tunnel from foundation pit reinforcement to excavation-backfilling

and layered excavation of the soil. The soil layers are modeled using the HSS model, with model parameters determined based on the project's geological survey report. The tunnel segments are modeled using a beam-hinge joint model.

Figures 5.62 and 5.63 show the tunnel segment deformation caused by the excavation of the soil above the up-line tunnel in Section N01 and the down-line tunnel in Section N02, respectively. From Fig. 5.62, it can be observed that excavation of the soil in Section N01 causes a heave of approximately 9.8 mm on the up-line, with a corresponding effect on the down-line, which experiences a heave of about 5.0 mm. According to the monitoring results shown in Fig. 5.57, during the excavation and

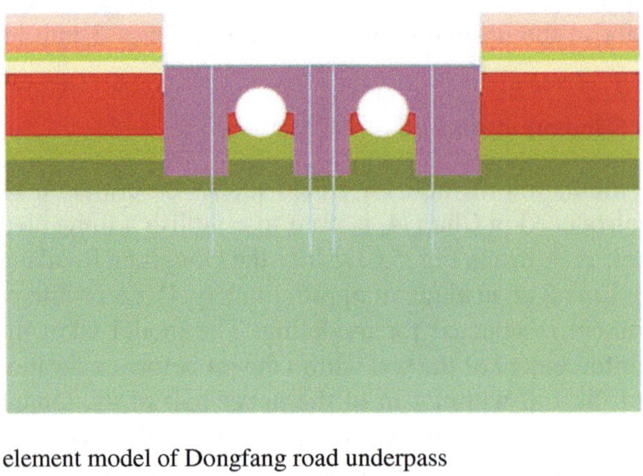

Fig. 5.61 Finite element model of Dongfang road underpass

backfilling of the soil above Section N01 on the up-line, the maximum heave of the up-line tunnel on Metro Line 2 reaches approximately 10.0 mm, with a heave of about 3.5 mm on the down-line. The simulation calculation results are in close agreement with the actual measurements.

When considering the overall deformation of the down-line tunnel segments, the impact on the down-line tunnel segments during the excavation of the soil above the up-line tunnel shows a rigid body displacement pattern following the deformation of the surrounding soil. The calculation results for the excavation of the soil above the down-line tunnel, shown in Fig. 5.63, also exhibit this deformation pattern for the up-line tunnel.

According to the deformation monitoring results for the down-line in Fig. 5.60, during the excavation of the soil above Section N02 of the down-line, the maximum heave of the down-line tunnel was approximately 6.0 mm. At this point, the concrete pouring for the middle partition wall and side walls in Section N01 had been completed, but the backfilling conditions were not considered in the calculation.

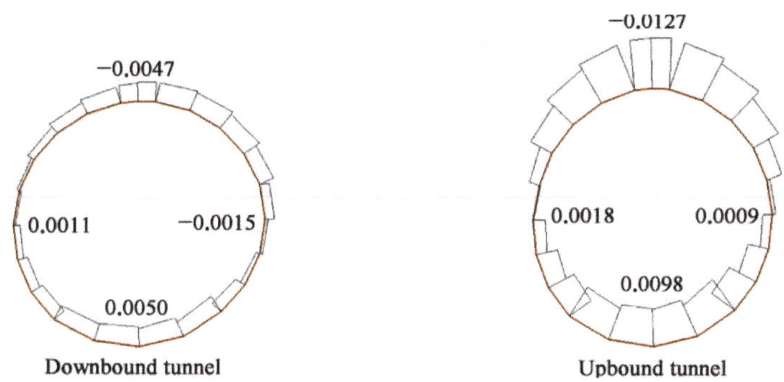

Fig. 5.62 Segment deformation induced by N01 zone excavation above upbound tunnel (unit: m)

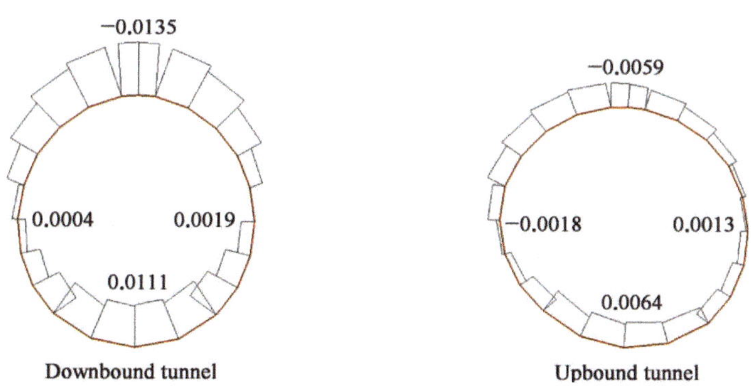

Fig. 5.63 Segment deformation induced by N02 zone excavation above downbound tunnel (unit: m)

Table 5.10 Calculated deformation statistics of tunnels in Dongfang road underpass project

Line	Unloading ratio n	Vertical uplift s/mm		Horizontal convergence Δd/mm		Uplift-depth ratio/‰	
		Measured	Calculated	Measured	Calculated	Measured	Calculated
Upbound tunnel (N01 zone soil excavation)	0.70	+10.0	+9.8		−2.7	1.59	1.48
Downbound tunnel (N02 zone soil excavation)	0.70	+5.0	+11.1		−2.3	0.80	1.68

Therefore, the excavation calculation results for the soil above Section N02 are slightly larger than the monitored results.

Table 5.10 lists the tunnel segment deformation caused by the excavation of the soil directly above the tunnel for the Dongfang Road underpass project. According to the calculation results, when the unloading ratio is 0.7, the tunnel heave is approximately 1.6‰ of the excavation depth.

The engineering case also clearly shows that the reinforcement process has a significant impact on the tunnel, whereas during pure excavation, the tunnel heave is only about 1‰ of the excavation depth. Therefore, it is essential to control the compressive disturbance during the reinforcement process.

5.4 Summary

This chapter presents five cases of tunnel deformation resulting from loading and unloading above the tunnel. The deformation patterns of the tunnels under these conditions were identified, and the control and remediation measures implemented in each case were outlined. The key deformation patterns and corresponding control measures are as follows:

(1) Under the effect of surcharge loading, different soil combinations exhibited different deformation characteristics: ① in the soft upper and hard lower soil combination, the difference in the horizontal diameter of the tunnel rings compared to the design value (ΔD) significantly increased, showing a distinct "horizontal duck egg" deformation. The longitudinal joints between the crown segment and the adjacent segment, as well as between the standard segment and the invert segment, experienced tensile opening, or cracks appeared in the concrete near the longitudinal joints. The longitudinal joints inside the standard segment and adjacent segment were compressed and closed, or the concrete protective layer was compressed and peeled off. ② in the hard upper and soft lower soil combination, when the soil below the tunnel within twice the tunnel

diameter is low in strength (such as mud), the tunnel may primarily experience settlement deformation compared to the relative convergence deformation.

(2) When the external longitudinal joints between adjacent and standard blocks open, the failure of the external water-stop strip results in leakage. Consequently, after large-scale loading in water-rich areas, continuous leakage frequently occurs at the tunnel's waist longitudinal joints. In contrast, due to the compression of the tunnel's top and bottom external walls, leakage problems at these locations are relatively rare during the initial stages of loading or when the loading effect is not severe.

(3) When tunnel damage occurs due to surcharge loading, the first step is to immediately initiate unloading from above. Subsequently, the leakage issue should be addressed by timely sealing to prevent surrounding pressure instability due to water and soil loss, which could exacerbate tunnel damage. If significant deformations or structural damage have already occurred, structural reinforcement measures—such as grouting corrections on the tunnel's exterior to restore surrounding pressure, or applying aramid fabric or steel lining to the tunnel interior—should be implemented.

(4) In soft soil areas, the excavation of foundation pits above tunnels should follow the "spatiotemporal effect" theory, with excavation conducted according to the principles of "layered, sectioned, symmetrical, balanced, and time-limited" methods. Excavation should take place during the night when trains are not in operation, and the size of each section should be planned so that both excavation and base slab pouring are completed within the non-operating period. If unloading leads to significant tunnel heave, ballasting measures can be used after the base slab's initial set.

(5) The compressive effect of foundation reinforcement on the tunnel must be strictly controlled. The construction method and parameters should be carefully chosen, as they are crucial in controlling the total heave during the entire unloading process.

Chapter 6
Conclusions and Mitigation Measures

6.1 Conclusions

Large-scale surface loading/unloading significantly influence both the stable strata and the tunnels in equilibrium. Changes in the loads above the tunnel lead to changes in the stress and displacement fields around the tunnel, exacerbating existing structural diseases or causing new ones, which will threaten the structural safety of tunnels in severe cases.

The effects of loading/unloading above soft-soil tunnels have been studied in previous chapters based on theoretical analysis, numerical calculations, model tests, field monitoring, and other analytical methods. On one side, the influencing factors, including tunnel burial depth, loading magnitude, loading range, geological characteristics, and relative spatial relationship, on tunnel deformation and other diseases under surface loading are analyzed; On the other side, the study focuses on the influence of foundation pit excavation on tunnel deformation, by analyzing the influencing factors, such as foundation pit excavation dimensions, soil reinforcement, anti-uplift pile settings, or excavation sequence. The study has promoted further insight into the influence of loading on tunnel damage and deepened the understanding of the tunnel deformation patterns and risk assessment. In conclusion, the influence characteristics of overburden loading and unloading on tunnels and surrounding ground are as follows.

R. Wang et al., *Impact Analysis and Control Strategies of Overlying Loading/Unloading Effects on Soft Soil Shield Tunnels*, Key Technologies for Tunnel Construction under Complex Geological and Environmental Conditions,
https://doi.org/10.1007/978-981-95-1462-5_6

6.1.1 Effects of Surface Loading on Strata and Tunnels

(1) Effects of surface loading on strata

Large-scale ground activities, such as earthwork disposal, soil piling, river backfilling, landscaping, and building construction, create a surface loading on the strata. The loading disturbs soil within a certain range and alters the original stress state of the strata, leading to a redistribution of stress and displacement fields, as has been well demonstrated through theoretical analysis, numerical calculations, field tests, and engineering practices. Notable examples include the Huqiu Tower in China, the Leaning Tower of Pisa in Italy, the Transcona Grain Elevator in Canada in the early 20th century, and the Kansai International Airport in Japan during the 1980s. The Huqiu Tower, built in 961 during the Song Dynasty, stands 47.5 m tall, with a tilt of 2.78° and a top offset of 2.31 m from the centerline. The Leaning Tower of Pisa, constructed in 1173, has a height of 60 m. Differential settlement occurred after completion, with an average settlement of 2.0 m, a maximum settlement of 4 m, a tilt of 5.5°, and a top eccentricity of 5.27 m. The Transcona Grain Elevator, completed in 1913, collapsed the same year it was commissioned. The structure was 31 m tall, supported by a 0.61-m-thick reinforced concrete raft foundation. The Kansai International Airport, constructed on 33-m land reclamation in 1986 and started operation in 1994. The Airport was designed to experience a settlement range of 5.7–7.5 m, and it was predicted that primary consolidation would be completed in 20 years. However, the settlement had already reached 8.1 m in 1990, necessitating an additional filling of 3 m beyond the original design.

From the settlement or differential settlement of these structures, it is evident that the weight of the buildings (including their self-weight and service loads) is not matched with the bearing capacity of the underlying strata. For soft strata, such as those along the rivers and coastlines in Shanghai, the soil strength is even lower, leading to greater deformation when subjected to large-scale loading. A notable example of this is the significant settlement of the Shanghai Exhibition Center, built in the 1950s. The effects of varying loading scales, magnitudes, and spatial relationships on the strata follow the patterns below:

① The larger the loading area, and the greater the loading magnitude (or the loading height), the greater the impact on the strata. This impact extends to a greater depth and persists for a longer time.
② With a fixed loading area, as the loading magnitude increases, the impact on the strata becomes deeper and more significant.
③ With a fixed loading height, as the loading area increases, the impact on the strata becomes greater.
④ The impact of the loading on the strata decreases as the distance from the loading increases.
⑤ The greater the load and the longer the loading time, the greater the consolidation settlement of the strata, and the smaller the rebound amount that the strata can recover after unloading.

Therefore, large-scale, close-range, and long-duration loading has a significant impact on the natural consolidation of strata. When the loading exceeds the bearing capacity of the foundation, it will damage the soil structure and have a major impact on the stress and displacement fields, with these changes often being nonlinear. Loading not only greatly affects soil deformation but also causes prolonged consolidation of the strata. If shield tunneling construction is carried out before the soil has fully consolidated over a certain period, the potential risks are undoubtedly significant.

(2) Effects of surface loading on tunnel structures and damage phenomena

The main hazards of surface loading on tunnels are manifested in the transverse convergence deformation of lining rings and the settlement of the arch bottom and crown. The former includes the bending deformation of the tunnel segments and the relative dislocation, rotation, opening, or closing of the segment joints, while the latter refers to the overall displacement of lining rings within the stratum. The impact of different loading conditions (such as loading scale, magnitude, distance, duration, etc.) and ground conditions (such as soil mechanical properties) on tunnels varies. Generally, it results in varying degrees of longitudinal settlement and transverse convergence deformation, leading to new structural diseases or exacerbating existing ones. These issues may include segment and joint deformation, opening of segment joints, misalignment or crushing of segments, segment cracking or compression damage, and water leakage. In severe cases, this can threaten the structural safety of the tunnel, potentially necessitating its shutdown for reinforcement. The following patterns can be observed:

① Both settlement deformation and convergence deformation of the tunnel approximately increase linearly with the loading magnitude. Sudden localized loading on the surface tends to cause significant damage to the tunnel, and the damage is difficult to recover after unloading.
② Tunnel settlement and convergence deformation are closely related to the mechanical properties of the strata. For tunnels embedded in homogeneous strata, the harder the strata, the smaller the effect of extensive surface loading on the convergence and settlement values of the tunnel; conversely, the greater the effect.
③ For a certain loading area, the greater the loading height (magnitude), the greater the effect on tunnel deformation. For a certain loading height (amount), the larger the loading area, the greater the range of influence on the deformation of the tunnel. When the loading area is large and covers the tunnel area, it is equivalent to full loading, meaning the effect of the loading on the tunnel depends directly on the loading height.

(3) Effects of surface loading and spatial relationship

The spatial relationship between the loading distribution and the tunnel position is crucial. The closer the loading is to the tunnel, the greater the impact; the farther away, the smaller the influence. When the extent of surface loading covers the tunnel or the closer it is to the tunnel axis, the greater the effect on tunnel settlement and convergence, and the greater the hazards caused. The effect on tunnel deformation becomes significantly smaller when the loading edge distance exceeds the range of $\pi/4 + \phi/2$ from the longitudinal axis of the tunnel.

① As the loading eccentricity increases, the additional earth pressure gradually decreases.
② As the loading eccentricity increases, the tunnel's convergence deformation first decreases rapidly and then tends to stabilize, transitioning from transverse oval deformation to an inclined oval deformation.
③ Increasing the loading eccentricity can reduce the deformation of the tunnel joints. Loading directly above the tunnel primarily causes deformation at the left and right shoulder joints and the crown joint, exhibiting symmetry. Eccentric loading mainly induces deformation at the shoulder and crown joints corresponding to the load position.

(4) Effects of stratigraphic conditions

Stratigraphic conditions have a great influence on the convergence and settlement deformation of tunnels. Sedimentary strata along the rivers and coasts are often characterized by "soft on the crown and hard on the bottom", which corresponds to the different strata of the tunnels and their underlying layers, and the deformation characteristics of the tunnels induced by the loading also show great differences. If the tunnel section and the underlying strata have similar or the same mechanical properties at the corresponding depth, the tunnel deformation is characterized by longitudinal settlement; if the performance of the strata through which the tunnel passes is much worse than that at the tunnel bottom, the tunnel deformation is characterized by transverse convergence.

① When the tunnel is entirely located in the soft strata, such as completely through Shanghai's soft Layer ④, and the tunnel bottom rests on thicker soft strata, the tunnel deformation is dominated by longitudinal differential settlement, with minor convergence deformation; if the tunnel bottom is located in harder and larger strata, the convergence deformation under upper loading becomes very obvious.
② The softness and hardness of both the strata through which the tunnel passes and the underlying strata are sensitive to loading. Under large-area surface loading, the softer the homogeneous soft strata, the greater the convergence, and the harder the homogeneous ground, the smaller the convergence. When the tunnel passes through the stratum and its underlying soft strata are thicker, the convergence deformation is not obvious and the tunnel is dominated by settlement deformation. When the tunnel passes through a softer stratum but the underlying stratum is harder, the tunnel surface is characterized by transverse convergence deformation.
③ The softness and hardness of the underlying strata significantly influence the convergence and deformation of the tunnel. For shield tunnels, the crown settlement depends mainly on the stiffness of the underlying layer: the softer the underlying layer, the larger the settlement, the harder the underlying layer, the

smaller the settlement. The stiffness of the crossing layer basically does not affect the magnitude of the invert settlement.

(5) Effects of loading on convergence deformation

Under loading, the difference ΔD between the horizontal diameter and the design value of each tunnel ring in the combination of upper soft and lower hard strata increases significantly, showing obvious "transverse oval" deformation. The inner side of the longitudinal joints between the capping block and the neighboring block as well as between the standard block and the invert block experiences tensile opening or cracking in the adjacent concrete, while the inner side of longitudinal joints between the standard block and the neighboring block are pressurized to close or suffer from protective layer spalling. The longitudinal joints between the neighboring blocks and the standard blocks are open on the outside, and the failure of the outer water stops results in water seepage problems; therefore, continuous leakage along waist longitudinal joints often occurs in water-rich areas after large-scale loading above the tunnels.

(6) Effects of loading on longitudinal deformation

Due to the longitudinal structural characteristics of the tunnel, surface loading causes tunnel longitudinal deformation, which manifests as both bending and misalignment deformation. The theoretical research analysis and engineering monitoring analysis have well reflected the longitudinal deformation characteristics of the tunnel. The patterns are as follows.

① The tunnel burial depth or the thickness of the upper overburden, the loading range and magnitude, the soil modulus of the strata, and the initial state of the tunnel have a great influence on the tunnel longitudinal deformation, and the tunnel settlement increases with the increase of the loading magnitude.
② The elastic modulus of the tunnel-crossing strata and the underlying strata has a significant effect on tunnel settlement and stress. Increasing the elastic modulus of the underlying soil can significantly reduce tunnel settlement; conversely, decreasing it greatly increases settlement. When the tunnel underlies soft soil strata, longitudinal deformation under surface loading becomes more pronounced; when the tunnel is entirely in a harder stratum, the settlement is smaller. Therefore, to ensure the safe operation of tunnels, surface loading above tunnels shall be prohibited.
③ For subway tunnels with through joints in soft soils in Shanghai, the tunnel longitudinal deformation under surface loading is likely to exceed the code limit.
④ As the loading eccentricity increases, the tunnel longitudinal deformation under surface loading becomes smaller, and when the loading eccentricity is greater than two times the outer diameter of the tunnel, the effect of loading is small.
⑤ The longitudinal and circumferential stiffness of the tunnel itself is much greater than that of the surrounding soils. Therefore, settlement deformations in tunnels are generally less than surface settlement at the same elevation.

6.1.2 Effects of Surface Unloading on Subway Tunnels

For tunnels in equilibrium, the effect of unloading on heave deformation is also significant. The effect is greatly influenced by construction activities with different construction processes affecting tunnels in varying ways. Engineering monitoring shows that disorderly unloading has a large impact on tunnels, while orderly and controlled unloading has a small impact. However, the final impact depends on the unloading magnitude and the improvement of strata.

To examine the influence of construction parameters on tunnel deformation, numerical calculation, and analysis were carried out by establishing a finite element model to regulate various factors, such as the planar dimensions of the pit excavation, soil reinforcement improvement, setting of uplift piles, and excavation construction time. Then, the construction parameters are used to compare with the on-site monitoring date. The conclusions are as follows:

(1) Effects of unloading on strata.

Different from loading, unloading causes a rebound in strata, lifting the tunnels within the strata together. In the late 1990s, unloading research was carried out during the station construction of the Shanghai Metro Line 2. The effect of excavation on strata heave was monitored by embedding magnetic rings in the strata. In alignment with the excavation conditions, the heave amount below the base slab due to excavation unloading exceeded 70 mm over a typical two-month construction cycle from excavation to base slab completion.

(2) Effects of unloading on tunnels.

A case study showed that new river channel construction and unloading above the tunnel with an overburden of about 9.5 m led to tunnel heave. After soil excavation of 5.5 m, the tunnel heave was nearly 22 mm within a month, and the heave amount was about 4 ‰ of the unloading depth. The unloading resulted in tunnel heave, misalignment of segment joints, water leakage, and other diseases. Compared to strata, the longitudinal side of the tunnel featuring greater stiffness experiences less heave.

(3) Changes in tunnel properties.

When there is foundation pit excavation and uploading above the tunnel, tunnel deformation can be seen in two aspects: the overall heave deformation caused by soil rebound and cross-sectional distortion. During distortion, segments transform from a transverse oval (stretched transversely and compressed longitudinally) to a longitudinal oval (stretched longitudinally and compressed transversely). Since the tunnel is an assembled structure, heave (or settlement) and convergence deformation do not occur independently. The diameter changes in both directions in diameters reflect a self-regulation mechanism of the tunnel to accommodate soil stress release. And the tunnel structural changes can be balanced and coordinated within a certain range.

(4) As the unloading ratio of overlying strata increases, the tunnel heave deformation also increases.

When the width of foundation pit excavation $W <$ the tunnel's outer diameter D, the increase in deformation is limited; when $W > D$ and $n > 0.5$, the increase in heave deformation is significant. Reinforcing the surrounding strata of the tunnel can inhibit the development of transverse convergence deformation of the tunnel and improve the deformation behavior of the tunnel. Different reinforcement methods have different impacts on the tunnel. Specifically, the higher the reinforcement strength and the larger the reinforcement range, the more effective the resistance to strata heave, leading to less tunnel deformation. Meanwhile, a close-range reinforcement process will also cause extrusion deformation of the tunnel.

(5) Full soil reinforcement of the foundation pit above the tunnel can effectively limit soil rebound.

For reinforced strata ($q_u \geq 0.8$ MPa), compared with unreinforced strata, the heave caused by foundation pit excavation above the tunnel is reduced by 50%, and the transverse convergence deformation Δd is also reduced by more than 50%. Therefore, foundation reinforcement is very effective in controlling tunnel heave deformation and convergence deformation induced by foundation pit excavation above the tunnel.

(6) The integration of pile foundation and base slab anchoring achieves notable effects in resisting heave deformation.

The deep pile is intrinsically effective in resisting foundation heave. Therefore, it can be integrated with the base slab to resist heave, reducing tunnel heave and convergence deformation by approximately 25%. Foundation pit excavation above the tunnel can control the ratio of the tunnel heave s to foundation pit excavation depth h falls within 0.78–1.71‰, with an average value of 1.16‰.

(7) Controlling the timing of foundation pit excavation helps control the tunnel deformation.

The longer the excavation time is, the greater the tunnel deformation is. The excavation time has a greater influence on tunnel heave deformation than transverse convergence deformation. The tendency of tunnel convergence deformation with time is influenced by the soil consolidation after soil disturbance.

When foundation pit excavation above the tunnel occurs, tunnel deformation is closely influenced by foundation pit size and retaining structure, soil reinforcement range and strength, excavation timing, etc. The control of tunnel deformation is not independent. It's necessary to consider a practical engineering environment when implementing deformation mitigation measures.

6.2 Mitigation Measures

(1) Principles of loading management

① When loading is detected, operators should promptly assess the surface surcharge conditions, and compare the original surface map to detail the topographic situation. The first step is to organize immediate unloading. Unloading the surface loads and reducing longitudinal loads can effectively improve the tunnel stress distribution and partially restore tunnel deformation, which is very favorable for the tunnel safety. Meanwhile, address the tunnel diseases.
② Strengthen the monitoring and inspection of the tunnel. Then, establish a corresponding analysis of tunnel deformation, loading, and unloading conditions to prepare for subsequent reinforcement management.
③ Formulate an unloading plan and implement unloading quickly. The principle is "unloading in layers and removal in time". Minimize the stratum and structural changes of tunnels caused by loading.
④ Based on the monitoring, testing, and inspection of internal tunnel diseases (water leakage, segment cracking, tunnel convergence, and settlement, etc.), conduct timely safety assessment.
⑤ Water leakage shall be addressed immediately to avoid soil erosion which leads to confining pressure instability and subsequently worsens tunnel diseases.
⑥ Rapidly mark the affected areas, and observe changes and trends.
⑦ Finally, when significant deformation or structural damage has occurred, external grouting correction around the tunnel is required along with steel lining installation and other reinforcement measures.

(2) Principles of unloading management

Unloading causes the rebound of the foundation, resulting in tunnel heave. When unloading above the tunnel is considered, many factors should be considered, such as the current condition of the tunnel, excavation depth, excavation and unloading magnitude, foundation reinforcement, retaining structure, pile foundation works, excavation and construction requirements. During excavation construction, the "spatio-temporal effect" theory should be followed. Based on the principle of "stratification, segmentation, symmetry, balance, and time limits", excavate rapidly, install supports, and construct base slabs. And rebuild quickly to increase the load, which can balance part of the unloaded weight. Conduct construction during the train suspension period at night. The block size for excavation is determined by the completion of soil excavation and the placement of base slabs during this period. When the unloading ratio causes significant tunnel heave, loading measures can be adopted after the initial consolidation of the base slabs.

(3) Management and control of significant tunnel deformation

The most common disease caused by loading in tunnels is the "transverse oval-shaped" deformation. After thoughtful research and analysis, a complete set of "external control and internal treatment" is introduced. The treatment idea of "unloading from above, correction laterally, and reinforcement internally" pioneered by Shanghai has become a model for tunnel deformation treatment, which stabilizes the tunnel structural state and protects the tunnel structural safety.

① Rapidly unload the upper loading.
② Conduct grouting correction for significant tunnel deformation to improve the tunnel conditions and tighten the bolts.
③ Carry out post-wall grouting for sections that have cracks and water leakage, and reinforce the internal tensile steel rings when lining rings experience significant deformation, improving the bearing capacity of the tunnel.

Unloading above the tunnel also threatens the structural safety of the tunnel. Unloading activities, such as river excavation or foundation pit construction will disturb the stable strata and tunnels in equilibrium, resulting in strata and tunnel heave. The changes in the tunnel stress state lead to complex structural deformation. And construction disturbances of varying degrees exert different impacts on the tunnel. Assuming that excavation magnitude and construction conditions are the same, the influence of foundation pit excavation is more pronounced in soft soils, while it is much smaller in hard strata. Unloading above the tunnel shall be conducted in a planned and orderly manner; otherwise, it may lead to severe damage to the tunnel. The 20 successful unloading engineering cases are highly instructive, contributing to an agreed practice of controlling tunnel deformation and meeting tunnel protection standards. Loading and unloading must be stopped and corrected when they endanger the tunnel structure.

MIX
Papier aus verantwortungsvollen Quellen
Paper from responsible sources
FSC® C105338

If you have any concerns about our products,
you can contact us on
ProductSafety@springernature.com

In case Publisher is established outside the EU,
the EU authorized representative is:
Springer Nature Customer Service Center GmbH
Europaplatz 3, 69115 Heidelberg, Germany

Printed by Libri Plureos GmbH
in Hamburg, Germany